RECONFIGURABLE COMPUTING

Reconfigurable Computing

Accelerating Computation
with Field-Programmable Gate Arrays

by

MAYA GOKHALE

Los Alamos National Laboratory,
NM, U.S.A.

and

PAUL S. GRAHAM

Los Alamos, NM, U.S.A.

 Springer

A C.I.P. Catalogue record for this book is available from the Library of Congress.

ISBN-10 0-387-26105-2 (HB)
ISBN-13 978-0-387-26105-8 (HB)
ISBN-10 0-387-26106-0 (e-book)
ISBN-13 978-0-387-26106-5 (e-book)

Published by Springer,
P.O. Box 17, 3300 AA Dordrecht, The Netherlands.

www.springeronline.com

Printed on acid-free paper

Printed in the Netherlands.

Contents

Acknowledgments

We would like to recognize the International, Space, and Response (ISR) Technologies Division and the Laboratory-Directed Research and Development (LDRD) Program at Los Alamos National Laboratory for their invaluable support during the writing and editing of this book.

We would like to acknowledge the contributions of two invited chapter authors. Reid B. Porter from Los Alamos National Laboratory wrote Chapter 6, *Image Processing*, providing an excellent discussion of how reconfigurable computing has been employed in the broad field of image processing. Dominique Lavenier and Mathieu Giraud from IRISA, Rennes France wrote Chapter 8, *Bioinformatics Applications*, drawing on their extensive background in bioinformatics to describe several applications from the field and the role of reconfigurable computing in these applications.

The material in Chapter 9, *Supercomputing Applications*, is derived from two papers written by researchers at Los Alamos National Laboratory. The first paper, "Accelerating Monte Carlo Radiative Heat Transfer Simulation on a Reconfigurable Computer: An Evaluation", was written by Maya Gokhale, Janette Frigo, Christine Ahrens, Justin L. Tripp and Ronald G. Minnich and was published in the *Proceedings of the 2004 International Conference on Field-Programmable Logic and Applications*. The second paper, "Acceleration of Traffic Simulation on Reconfigurable Hardware", was written by Justin L. Tripp, Henning S. Mortveit, Matthew S. Nassr, Anders A. Hansson, and Maya Gokhale and was presented at the 2004 International Conference on Military and Aerospace Programmable Logic Devices.

Thanks are due to Janette Frigo for the phase modulation sorter example and to Kris Gaj and Peter Bellows for very helpful discussions on cryptography and network security.

We would also like to acknowledge the support from our families during the pursuit of this endeavor. The importance of their support during this project cannot be overstated. Of special note, Ron Minnich gave up many hours reading through the manuscript and providing us valuable feedback while Samara Graham took on the extra duty of trying to keep three children in line while Paul worked many late evenings.

This book is dedicated to our families.

1

An Introduction to Reconfigurable Computing

Reconfigurable Computing (RC), the use of programmable logic to accelerate computation, arose in the late '80's with the widespread commercial availability of Field-Programmable Gate Arrays (FPGAs). The innovative development of FPGAs whose configuration could be re-programmed an unlimited number of times spurred the invention of a new field in which many different hardware algorithms could execute, in turn, on a single device, just as many different software algorithms can run on a conventional processor.

The speed advantage of direct hardware execution on the FPGA – routinely 10X to 100X the equivalent software algorithm – attracted the attention of the supercomputing community as well as Digital Signal Processing (DSP) systems developers. RC researchers found that FPGAs offer significant advantages over microprocessors and DSPs for high performance, low volume applications, particularly for applications that can exploit customized bit widths and massive instruction-level parallelism. An even more compelling argument for using FPGAs as reconfigurable computers has been the commercial availability of devices that continue to track Moore's Law. FPGAs contain large amounts of SRAM combined with regularly tiled and interconnected logic blocks. These devices follow the International Technology Roadmap for Semiconductors (ITRS) roadmap [3] for memory rather than microprocessors and are often first on the leading new fabrication lines. Thus reconfigurable computers advance technologically at a faster rate than microprocessors.

1.1 What is RC?

The speed advantage of FPGAs derives from the fact that the programmable hardware is customized to a particular algorithm. Thus the FPGA can be configured to contain exactly and only those operations that appear in the algorithm. In contrast, the design of a fixed instruction set processor must accommodate all possible operations that an algorithm might require for all possible data types. An FPGA can be configured to perform arbitrary fixed precision arithmetic, with the number of arithmetic units, their type, and their interconnection uniquely defined by the algorithm. The fixed instruction processor contains ALUs of a specific width (16-, 32- or 64-bit) and has pre-determined control and data flow patterns. Reconfigurable vs. fixed processors are contrasted in Figure 1.1. In this example, the FPGA is configured to hold an array of application-specific processing units. Each processing unit contains four 8-bit adders and a 16-bit multiply-accumulate unit, all connected through registers. Hardware address generators are used to access off-chip data. The microprocessor in this example has a Harvard architecture. It accesses the sequential instruction stream through the Instruction Cache. The data memory hierarchy includes external memory, cache and integer and floating point register files, which supply operands to the arithmetic units [207].

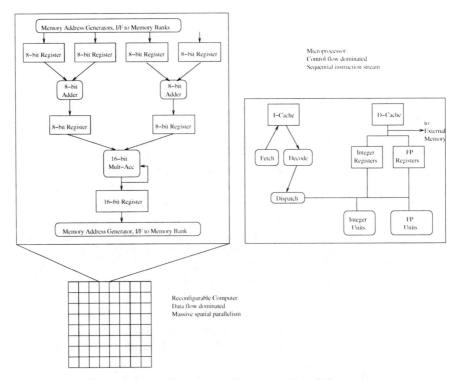

Fig. 1.1. Reconfigurable vs. Processor-Based Computing

The example illustrates the major differences between reconfigurable and processor-based computing. The FPGA is configured into a customized hardware implementation of the application. The hardware is usually *data path* driven, with minimal control flow; processor-based computing depends on a linear instruction stream including loops and branches. The reconfigurable computer's data path is usually pipelined so that all function units are in use every clock cycle. The microprocessor has the potential for multiple instructions per clock cycle (IPC) [207], but the delivered parallelism depends on the instruction mix of the specific program, and function units are often under-utilized. The reconfigurable computer can access many memory words in each clock cycle, and the memory addresses and access patterns can be optimized for the application. The processor reads data through the data cache, and efficiency of the processor is determined by the degree to which data is available in the cache when needed by an instruction. The programmer only indirectly controls the cache-friendliness of the algorithm, as access to the data cache is hidden from the instruction set architecture. To summarize, reconfigurable computing is concerned with decomposing applications into spatially parallel, tiled, application-specific pipelines, whereas the traditional general purpose processor interprets a linear sequence of instruction, with pipelining and other forms of spatial parallelism hidden within the microarchitecture of the processor.

1.2 RC Architectures

FPGAs form the processing building blocks of reconfigurable computers, as shown in Figure 1.2. The most common RC configuration is an accelerator board that plugs into the I/O slot of a microprocessor. The plug-in board typically contains

- one or more FPGAs,
- interface logic for the FPGAs to communicate with the conventional computer's I/O bus,
- memory local to the RC board, often double or quad data rate (DDR or QDR) Static Random Access Memory (SRAM) and/or higher capacity Synchronous Dynamic RAM (SDRAM),
- A/D interfaces or other serial communication links to acquire data or communicate over a network.

As an accelerator, the RC is used in one of two scenarios. The FPGAs can be used to process high bandwidth data streams from the external serial I/O interface, perform data reduction, and send processed data at a lower volume to the host processor. Alternatively, the host sends archival data to the RC memory subsystem. The FPGAs perform compute-intensive processing of the on-board data, and write results back into local memory, which the host then retrieves.

A second RC scenario is as an acceleration component of a cluster supercomputer. In this configuration, the I/O interface on the FPGA board communicates with the interconnection network of the supercomputer, and gives the board access to the supercomputer's global memory address space. An FPGA communicates over the interconnection network with processors

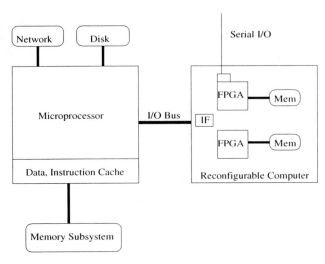

Fig. 1.2. Reconfigurable Computing System

and other FPGAs. The FPGAs may themselves be connected to each other via a private FPGA-only network through the on-chip high speed serial I/O.

1.3 How did RC originate?

While the basic notion of reconfigurable computing appeared in the 1960's [140], RC originated in practical terms in the late 1980's with the emergence of Field-Programmable Gate Arrays (FPGAs), Integrated Circuits whose hardware personality could be completely re-defined simply by loading a new "configuration," just as new software modules can be loaded onto a microprocessor or DSP. Mapping data- and compute-intensive algorithms to FPGAs could yield the speed approaching Application Specific Integrated Circuits (ASICs) with the flexibility of software.

Researchers in the United States and France in search of flexible, high performance building blocks, envisioned a new kind of supercomputer, composed of hardware-re-programmable components, that, by customizing the hardware to each application in turn, could deliver one to two orders of magnitude performance increase over convention fixed instruction set processors. The first reconfigurable computers were built by the IDA Supercomputing Research Center (SRC, re-named Center for Computing Sciences in 1994) in the USA and the DEC Paris Research Lab (closed after Digital Equipment Corporation was sold).

Two versions of the "Splash" systolic array were built at the SRC. The original Splash board, built in 1989 and costing about $13,000 in parts, could outperform an extant Cray 2 supercomputer for bit-oriented linear pattern matching applications [168]. The system contained 32 Xilinx 3090 series FP-GAs connected in a linear array. Adjacent FPGA chips shared a memory buffer. The RC was connected to a Sun workstation via VME interconnect. Splash 1 could perform DNA sequence comparison at 45X a (1990 era) high performance workstation.

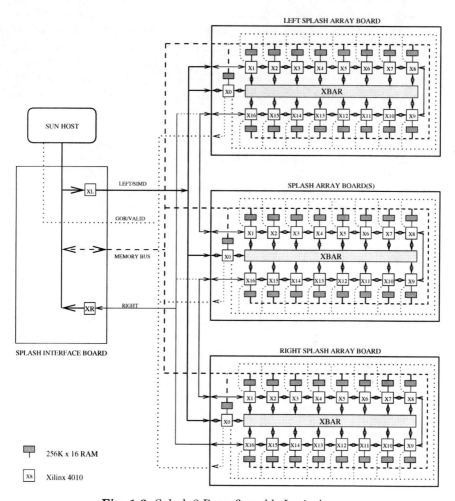

Fig. 1.3. Splash 2 Reconfigurable Logic Array

Splash 2 [60], built three years later, reduced the number of FPGAs to 16 (see Figure 1.3. However, due to rapidly increasing density of FPGAs, Splash 2, with 16 Xilinx 4010 FPGAs contained $1\frac{1}{2}$ times the logic of Splash 1. To improve interconnect flexibility, Splash 2 augmented the linear interconnect with a crossbar, allowing any FPGA to communicate directly with any other.

To improve scalability, boards could be linearly interconnected, providing for up to 8 boards. Rather than require the FPGAs to arbitrate a shared memory buffer, each FPGA had access to a dedicated SRAM module. The I/O interface to the host was updated to SBUS.

Concurrently with the Splash board development, the DEC PRL built a "Programmable Active Memory" (PAM) Perle-0 [47] .Targeted toward image processing applications, this system contained a 5×5 mesh of Xilinx 3020 FP-GAs with a VME interface to a host processor. Applications such as 512-bit multiplication, data compression, and image processing were benchmarked on this early board, demonstrating speed-ups of 2–100 compared to software implementations. Four generations of PAM boards were built, updating the FP-GAs and the host interconnect as with the Splash series [48], and many novel and high performance applications have been demonstrated on the Pamette, the PAM successor, e.g., [364], [365].

Following the success of these early research machines, FPGA-based reconfigurable computers became commercialized, and today there are dozens of RC accelerator board vendors as well as many custom RC boards being fabricated. In Chapter 3 we will discuss the architecture of modern reconfigurable computers in depth.

1.4 Inside the FPGA

The Field-Programmable Gate Array (FPGA) is the computational unit for RC systems. The FPGA is a regularly tiled two-dimensional array of logic blocks. Each logic block is a Look-Up Table (LUT), a simple memory that can store an arbitrary n-input boolean function. The logic blocks communicate through a programmable interconnection network that includes both nearest neighbor as well as hierarchical and long path wires. The periphery of the FPGA contains "I/O blocks" to interface between the internal logic blocks and the I/O pins. This simple, homogeneous architecture has evolved to become much more heterogeneous, including on-chip memory blocks as well as DSP blocks such as multiply/multiply-accumulate units.

While there are several sorts of FPGAs, include those that can be programmed only once, reconfigurable computers have exclusively used SRAM-programmable devices. This means that the configuration of the FPGA, the "object code" defining the algorithm loaded onto the device, is stored in an on-chip SRAM. By loading different configurations into the SRAM, different algorithms can be executed. The configuration determines the boolean function computed by each logic block and the interconnection pattern between logic and I/O blocks. The internal FPGA structures is described in Chapter 2.

1.5 Mapping Algorithms to Hardware

The task of mapping an algorithm onto a collection of configurable logic blocks is daunting indeed. FPGAs were invented for hardware prototyping and "glue logic" functions that are in the realm of hardware engineers who create an optimized design once and then use it for the lifetime of a product. It is desirable to create a multitude of algorithms and to continually modify and update them.

Unfortunately, FPGA languages and design tools are derived from the Application Specific Integrated Circuit (ASIC) design flow. Hardware design languages (HDL) or schematic entry tools are commonly used to create RC configurations. The detailed clock cycle level behavior of the circuits must be specified. The designer must trade off combinational vs. sequential circuits, design data paths and state machines to control them, and handle I/O to memory and external devices. The memory hierarchy is completely exposed, so the designer must choose among levels of memory hierarchy (external memory, on-chip memory, logic blocks configured as memory, register) for logical program variables and specify their access in the HDL. The timing behavior of each sort of memory must be explicitly factored into the circuit, and any changes, for example from external memory to on-chip memory, may result in changes throughout the design. Application specific modules implementing the desired algorithm must interface to vendor-supplied or third party Intellectual Property (IP) modules, e.g., to communicate over Ethernet or access the PCI bus. Using HDL, pipelines must be manually constructed.

Research into High Level Languages (HLL) for reconfigurable computing has been actively pursued. Emerging C and Java compilers offer a much higher level of abstraction than HDL or schematic entry. However, quality of compiler-generated circuits still lags the equivalent manual designs, often by factor of 2–4 in area.

Compiling an algorithm description (either in HDL or HLL) to an FPGA is also a complex task. Rather than a simple push button compile process, measured in seconds or minutes, there is a long, complex tool chain whose ultimate output, the configuration bit stream, may take hours to generate. The very flexibility that make FPGAs universal hardware also create significant challenges for the tools that map algorithm to logic blocks and routes.

In software, an algorithm is described in an abstraction of instruction set architecture, and through compilation is mapped to a specific instruction set. Not only does the algorithm description language closely model the processor architecture, but many design choices are hidden from the software programmer and the instruction set architecture. For example, virtual memory hardware, cache, number of function units are design parameters determined within by the processor's micro-architecture, and are not accessible to the programmer. On a reconfigurable computer, low level choices are available to the synthesis tool. HLL (behavioral) compilers must select memory hierarchy, synthesize datapath and control, and allocate function units. HDL compilers must perform logic synthesis from HDL descriptions. FPGA-specific tools then map gate-level descriptions onto configurable logic blocks and perform placement and routing. With a very large search space, optimization techniques such as simulated annealing are employed in the place&route stage.

Once an algorithm has been converted to hardware, debugging it requires the engineer to understand the low level parallel behavior of the generated circuits, at the granularity of nanoseconds. The fine grained spatial parallelism of hardware offers many opportunities for errors through race conditions, leading to non-deterministic behavior and even deadlock or livelock. Debugging the hardware representation of an algorithm is intrinsically more difficult than the corresponding software. Compounding the problem is lack of visibility into hardware state on the FPGA.

The success of reconfigurable computers ultimately hinges on how these difficult software development challenges can be solved. Languages, compilers, and debug tools for RC are discussed in Chapter 4.

1.6 RC Applications

Despite the formidable challenges associated with converting algorithms to hardware, reconfigurable computing has flourished as a discipline. Interest in RC is spurred by

- the 10–100X speedup demonstrated for certain data intensive application classes,
- compact form factor of reconfigurable computing systems, and
- non-RC commercial drivers for continual improvement of FPGAs.

The earliest RC applications were signal and image processing. With FPGAs, a significant fraction of the signal processing pipeline, from A/D input through FFTs and event detection, can be mapped onto configurable hardware and process hundreds of mega-samples per second (Chapter 5). Similarly, using pipelining and spatially tiled parallelism, RC image processing applications can maintain real-time video frame processing rates, as discussed in Chapter 6. Network security applications, in which the reconfigurable processor can operate at the host, gateway, or router, similarly offer line rate processing of encryption/decryption algorithms, packet filtering, and network intrusion detection. These applications, discussed in Chapter 7, exploit RC pattern match algorithms, as do bio-informatics applications, the subject of Chapter 8. Recently, RC has even entered the realm of supercomputing (Chapter 9) as floating point computation can be performed on large, high performance FPGAs.

1.7 Example: Dot Product

A simple dot product operator serves as an excellent illustration of reconfigurable computing applications. Dot product is the kernel operation in matrix multiply as well as image processing convolution operators. The dot product kernel is shown in Figure 1.4.

```
int acc;
int coeff[n], data[n];
for (i=0; i<n; i++)
    acc += coeff[i] * data[i];
```

Fig. 1.4. Dot Product

On conventional computer systems, the simple C code is sufficient to compile and obtain efficient execution. On a reconfigurable computer, while compilers do exist to synthesize hardware from the C code, significantly more efficient hardware can be obtained through manual techniques. With a microprocessor, the data type is simply "int." On the RC, the multiply accumulate unit, registers, and memory could be configured to several different sizes – from 8-bit to 64-bit – with very different requirements in area on chip. On a conventional computer, the memory hierarchy is managed within the microarchitecture of the processor, so that loading array elements from main memory into the cache is transparent to the compiler. On the RC, data arrays could be mapped among different off-chip memory banks in blocks or be interleaved. Portions of the arrays might even be cached through on-chip RAMs. Design choices on the arithmetic units include

- selection of data sizes – for example, the designer might choose 8-bit registers, a 16-bit multiplier and 16-bit accumulator
- selection of Intellectual Property (IP) blocks for the adders and multiplier – "soft" IP cores mapped onto configurable logic blocks may be selected, or alternatively, if available, "hard" multipliers and multiply-accumulate units may be selected, if available on the FPGA

The loop must be separated into data path and control, which can be done by behavioral synthesis tools. For optimal performance, the loop can be pipelined, which implies pipelining memory accesses, inserting delay registers, and generating pipeline control. The size of the data arrays and sizes/number of external memories, all affect the form of the loop pipeline, and the large design search space stretches the capabilities of RC compilers. Finally, the kernel application must interact with other components such as memory or microprocessor, all of which require that the designer understand the interface characteristics and timing.

If the design has been manually created, the next step is simulation using a test bench to insert test vectors into the design and observe waveforms. The designer is not only debugging the core user application, but also the interaction with external "black box" components such as memory, processor bus, I/O device or interconnection network.

After the design behaves correctly in simulation, it can be synthesized, placed, and routed, as outlined above. The generated configuration bit stream is, however, only part of the overall application. RC systems are typically controlled by conventional computers (a "host"). Therefore, a host program must be created to load the configuration, pre-load RC memories, start the RC, monitor progress, and retrieve results.

As can be seen from this scenario, successfully mapping an application onto a RC system to obtain the 10–100X speedup requires a concomitant 10X in effort.

1.8 Further Reading

While this volume is one of the first comprehensive surveys of reconfigurable computing with FPGAs, there are many RC resources to explore. [85] provides a concise survey of configurable computing systems, with an emphasis on architecture and compilation tools. In [400], the theory and algorithms for dynamic reconfiguration are discussed. [60] describes the Splash 2 reconfigurable computer. The volume [288] concerns signal processing algorithms and their implementation in programmable hardware. For more general texts on FPGAs, [419] (for system design using FPGAs) and [59] (for FPGA architecture) are excellent general references.

There are several conferences and workshops devoted to reconfigurable computing. The International Conference on Field-Programmable Logic and Applications (FPL), started in 1991, is the oldest reconfigurable computing conference, followed close behind by the IEEE Symposium on Field-programmable Custom Computing Machines (FCCM) in 1993 and the ACM International Symposium on Field-Programmable Gate Arrays (FPGA) in 1995. The Military and Aerospace Programmable Logic Devices International Conference (MAPLD), sponsored by NASA, has been held since 1998. Other conferences and workshops are the International Conference on Field-Programmable Technology (FPT), the International Conference on Engineering of Reconfigurable Systems and Applications (ERSA), held in conjunction with a computer science Multi-conference, and the Reconfigurable Architectures Workshop (RAW), associated with the International Conference on Parallel and Distributed Systems. All of these conferences provide a wealth of original papers on all aspects of reconfigurable computing.

Reconfigurable Logic Devices

Though Gerald Estrin [139, 141, 142] had conceived of reconfigurable computing as early as 1960, the relatively recent developments in reconfigurable computing have been fueled by the availability of logic devices that can be quickly and easily programmed and reprogrammed to perform a large variety of functions. The first devices of this type that achieved enough density to perform significant portions of a computation and that had significant availability were field-programmable gate arrays (FPGAs). These chips provide the designer with arrays of simple logic functions and memories (such as flip-flops) that can be connected through programmable interconnection networks. To begin with, the early FPGA devices from Xilinx, Altera, and others provided relatively little logic, but later generations provided enough logic for researchers to consider what might be possible through direct implementation of computational algorithms in reconfigurable logic devices. The densities of today's FPGAs have exceeded 150,000 4-input look-up tables (LUTs) per device and some have developed into devices that can be used to build complete systems on a programmable chip (SoPC), providing such specialized features as digital signal processing (DSP) blocks, multi-gigabit serial I/O, embedded microprocessors, and embedded SRAM blocks of various sizes.

In addition to the relatively fine-grained configurability provided by FPGAs and similar devices, the drive to reduce the power, area, and/or delay costs of fine-grained reconfigurability has led to a number of what may be called "coarse-grained" reconfigurable logic devices. Instead of providing configurability at the level of individual gates, flip-flops or look-up tables (LUTs), these coarse-grained architectures often provide arithmetic logic units (ALUs) and other larger functions that can be combined to perform computations. In the extreme, the functions might be as large as microprocessor cores such as in the Raw chip [408] from MIT.

The goal of this chapter is to provide the reader with a brief overview of what reconfigurable logic is and some of the important forms of reconfigurable logic that have been used in reconfigurable computing. This chapter will discuss, in general terms, the features of both fine- and coarse-grained reconfigurable logic devices and will describe some significant examples of each.

2.1 Field-Programmable Gate Arrays

With their introduction in 1985, field-programmable gate arrays (FPGAs) have been an alternative for implementing digital logic in systems. To begin with, FPGAs were used to provide a denser solution for glue logic within systems, but now they have expanded their applications to the point that it is not uncommon to find FPGAs as the central processing devices within systems. Compared with application-specific integrated circuits (ASICs) and mask-programmable gate arrays (MPGAs), FPGAs have several advantages for their users, including: quick time to market, being a standard product; no non-recurring engineering costs for fabrication; pre-tested silicon for use by the designer; and reprogrammability, allowing designers to upgrade or change logic through in-system programming. By reconfiguring the device with a new circuit, design errors can be fixed, new features can be added, or the function of the hardware can be entirely retargeted to other applications. Of course, compared with ASICs and MPGAs, FPGAs cost more per chip to perform a particular function so they are not good for extremely high volumes. Also, an FPGA implementation of a function is slower than the fixed-silicon options. Over time, though, the expense of doing custom silicon and the fact that FPGAs now tend to use state-of-the-art CMOS processes mean that FPGAs are performing more of the functions that ASICs and MPGAs would have performed in many systems.

Considering the complexity of contemporary FPGA devices and the several existing texts describing FPGA architecture and design [50, 59, 395], this section will provide only a brief overview of FPGA architecture and features as a foundation for later discussions about how FPGAs are used in reconfigurable computing. This overview will first discuss basic FPGA architecture, including basic logic, routing, and input/output (I/O) structures. Then, we will introduce some of the more advanced features that have been integrated into FPGAs since the late 1990's, including embedded memory, embedded arithmetic logic, high-speed serial I/O, and embedded microprocessors. While introduced from the perspective of FPGAs in this chapter, tightly integrated reconfigurable logic and microprocessors will be discussed later in Chapter 3 in the context of reconfigurable computing systems. This section on FPGAs will conclude with a discussion on FPGA programming architecture.

2.1.1 Basic Architecture

The basic architecture of FPGAs consists of three kinds of components: logic blocks, routing, and input/output blocks. Generally, FPGAs consist of an array of programmable logic blocks that can be interconnected to each other as well as to the programmable I/O blocks through some sort of programmable routing architecture. Figure 2.1 provides a very simplified diagram of a generic FPGA architecture.

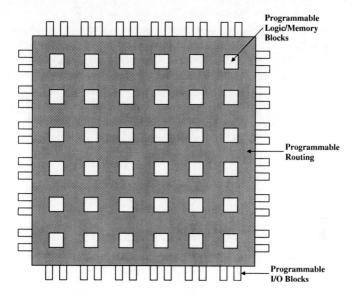

Fig. 2.1. A Generic FPGA Architecture

With fine-grained architectures such as FPGAs, the logic operations are mainly done at the bit or small word (≤ 4 bits) level, providing a very flexible architecture that can be customized to the specific needs of an application. For instance, if an application only needs a 6-bit adder for a particular operation and a 23-bit adder for another, the designer does not need to use a 32-bit adder to perform these lower precision computations—the designer can directly implement the size of adder that is needed for the specific application. This flexibility does comes at a significant cost compared with a non-programmable, custom silicon implementation due to the large number of transistors and the large amount of wiring needed to provide the fine-grained programmability. Numbers such as 10X-100X cost in terms of silicon area and 10-100X cost in circuit speed are frequently quoted [8] for FPGA implementations as compared to custom ASICs. Designers of FPGA chips try to balance the costs of flexibility with those of chip size and speed to give FPGA users both the flexibility they desire while reducing chip costs and increasing chip speeds. The next several paragraphs will describe the main three elements of FPGAs and some typical examples of how they are constructed.

Programmable Logic

FPGA designers have developed a large variety of programmable logic structures for FPGAs since their invention in the mid-1980's. For more than a decade, much of the programmable logic used in FPGAs can be generalized as shown in Figure 2.2. The basic logic element generally contains some form of programmable combinational logic, a flip-flop or latch, and some fast carry logic to reduce the area and delay costs for implementing carry logic. In our generic logic block, the output of the block is selectable between the output of the combinational logic or the output of the flip-flop. The figure also illustrates that some form of programming, or *configuration*, memory is used to control the output multiplexer; of course, configuration memory is used throughout the logic block to control the specific function of each element within the block.

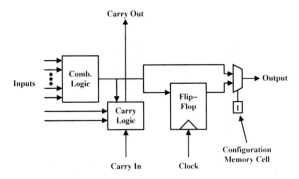

Fig. 2.2. A Generic Programmable Logic Block

Unlike our generic logic element, commercial FPGA devices generally provide a large amount of flexibility within the logic element. For instance, a flip-flop in many commercial FPGAs can be made to operate as a simple latch, can be programmed to have several combinations of asynchronous or synchronous sets and resets, and can be negative- or positive-edge triggered. In recent Xilinx and Altera devices, the carry logic has evolved into logic that supports additional functions. For instance, with the Xilinx Virtex FPGA, the carry logic has been augmented to help with multiplication. In the latest Stratix II FPGA from Altera [260], full-adders have taken the place of carry logic.

Regarding the combinational logic portion of the logic element, many different implementation methods have been used. The most common way to implement the combinational logic is with a look-up table (LUT). Figure 2.3 illustrates how a three-input look-up table is conceptually implemented—a series of programmable memory cells with a multiplexer to select the output of a specific memory cell. The look-up table, of course, operates as a memory with N address lines and 2^N memory locations. To implement a specific logic function with the memory, the truth table for the function is loaded into the memory. For example, the look-up table in Figure 2.3 implements an three-input AND function. Note that the LUT only produces a logic "1" when all address lines (labeled A, B, and C) are a logic "1"; this, of course, is due to the fact the memory cell at address 7 is the only cell storing a "1" value. Due to area efficiency [348], most commercial SRAM-based FPGAs use four-input LUTs for their combinational logic elements.

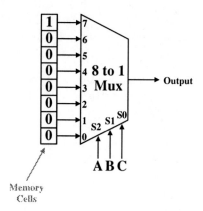

Fig. 2.3. A Three-Input Look-Up Table

Though most reprogrammable FPGAs use LUTs for combinational logic, several architectures (e.g., [5,162,234]) have used combinations of multiplexers and logic gates to implement programmable logic structures. An example of this approach can be seen in the Actel ProASIC Plus logic element shown in Figure 2.4, which can produce all three-input/one-output functions except for the three-input exclusive OR (XOR) and can even operate as a master-slave flip-flop of various types. Reprogrammable FPGAs using these alternative combinational logic architectures tend to implement functions of a finer granularity (2- and 3-input functions are common) at the logic-element level than LUT-based architectures.

To reduce the costs of using programmable routing, many reprogrammable FPGA architectures will cluster the logic elements together using fast, short-length routing. This clustering allows larger functions to be created using only the faster routing of the cluster. Most recent LUT-based architectures employ this strategy, often pairing two or more four-input LUT-based logic elements into a cluster. Over time, due to the delay costs of general-purpose programmable routing, commercial LUT-based FPGAs have increased the size of the clusters to improve circuit performance. Table 2.1 illustrates this increased clustering over time. In their Stratix II FPGA [15,260], Altera uses a mini-cluster they call the "Adaptive Logic Module" (ALM) consisting of four

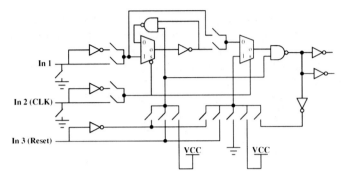

Fig. 2.4. Actel ProASIC Plus Logic Element

three-input LUTs and two four-input LUTs to create a flexible combinational logic structure that can handle up to a single function of 7 inputs or two independent functions of between 3 to 5 inputs. The full cluster in Stratix II, called a Logic Array Block (LAB), then collects 8 ALMs into a larger unit, resulting in a cluster of 24 three-input LUTs and 16 four-input LUTs.

Device Name	Year	LUT Width	Cluster Name	Cluster Size
Xilinx XC2000	1985	4	CLB	1
Xilinx XC3000	1987	4	CLB	2
Xilinx XC4000	1990	3 & 4	CLB	1 (3LUT) & 2 (4LUT)
Altera FLEX 8000	1992	4	LAB	8
Altera FLEX 10K	1995	4	LAB	8
Xilinx Virtex	1998	4	CLB	4
Altera Apex 20K	1998	4	LAB	10
Xilinx Virtex-II	2000	4	CLB	8
Altera Apex II	2001	4	LAB	10
Altera Stratix	2002	4	LAB	10
Xilinx Virtex-4	2004	4	CLB	8
Altera Stratix II	2004	3 & 4	LAB	24 (3LUT) & 16 (4LUT)

Table 2.1. Logic Element Cluster Sizes of LUT-Based FPGAs over Time

Routing

As with logic element structures, FPGA designers have used a variety of routing structures within their FPGAs. Various forms of routing exist through out each FPGA architecture. Generally, some amount of routing is included within each logic cluster so that the logic elements can be combined to form larger functions. External to the logic clusters is the more global routing architecture of the FPGA. In this section we will concentrate mainly on global routing architectures, though some general comments will be made about the internal cluster routing.

To implement programmable routing, three basic switch types are used: multiplexers, pass transistors, and tri-state buffers. Figure 2.5 illustrates each of these switches with an SRAM cell controlling their outputs. Generally, pass transistors and multiplexers are used within a logic cluster to connect the logic elements together while all three are used for the more global routing structure. Multiplexers are very common switch elements in FPGAs and they come in a large variety of widths ranging from two inputs to eight or more inputs, depending on the complexity of the routing network. Every time a signal is routed through pass-transistor-based switches, more capacitance is added to the load of the driving transistors. To counteract the resulting slow down due to capacitance, most modern FPGAs include active buffering in the routing networks.

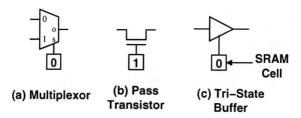

Fig. 2.5. Basic Programmable Switch Types

Within a logic cluster, routing is used for several purposes. First it is used to determine where the inputs to the logic elements come from and where the outputs will go. Next, programmable routing is used to determine how signals propagate through the logic elements themselves. Further, non-programmable routing is generally used for fast carry propagation to eliminate the extra delays incurred when using programmable routing. Usually, these carry chain paths extend between logic clusters as well, to support wide additions. Finally, routing within the logic cluster is frequently used to combine the logic elements into wider or otherwise more complex functions. Figure 2.6 illustrates different sorts of routing within a logic cluster.

Several global routing architectures have been implemented in FPGAs, the main four can be categorized as the island, cellular, long-line, and row architectures [49,395]. We will briefly describe each. Note that modern FPGAs have routing architectures which are significantly more complex than what we describe, but these general architectures are still identifiable within these FPGAs.

Figure 2.7 illustrates the basic island-style routing architecture. In this routing architecture, logic clusters are surrounded by segmented horizontal and vertical routing channels. Each cluster connects to the routing through "connection boxes" and each segment in the routing can be connected to an-

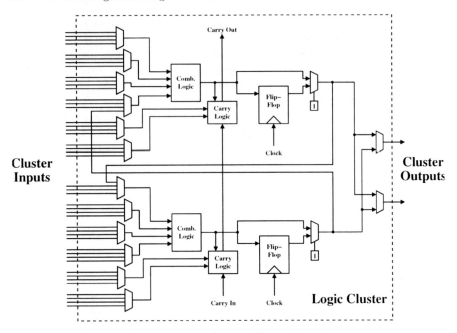

Fig. 2.6. Internal Logic Cluster Routing

other segment through a "switch box." The main feature of this type of routing is that connections between logic clusters are made through segmented routing. This architecture is found in many Xilinx FPGA architectures, though, Xilinx provides segments in several lengths (even wires that span the entire chip) while also providing local routing between logic clusters to make the architecture more efficient.

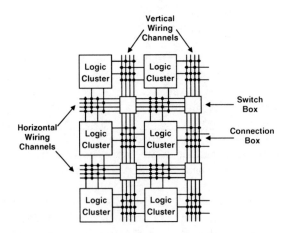

Fig. 2.7. Island Routing Architecture

The next architecture, the long-line routing architecture, takes a different approach. As illustrated in Figure 2.8, the long-line architecture also surrounds logic clusters with horizontal and vertical routing channels with multiple wires per channel, but each of the wires spans the width or height of the entire chip. Ideally, to connect any two logic clusters in this arrangement, only one vertical and one horizontal long line is required. The transition between a long vertical line and a long horizontal line can be made by using the internal routing of a logic cluster at the intersection of the two lines. This has been the main routing architecture for Altera FPGAs, but other FPGAs such as Actel's ProASIC FPGAs have similar routing structures. As illustrated in the figure, Altera's routing architecture generally provides for horizontal, local inter-cluster routing as well. To reduce the speed penalty for driving wires that span the length of the chip, several of Altera's latest architectures, such as Stratix [261] and Stratix II [15,260], have introduced smaller length segments in the routing channels.

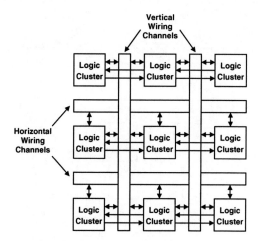

Fig. 2.8. Long-Line Routing Architecture

The cellular routing architecture, shown in Figure 2.9 is different yet from the island and long-line architectures. In this routing architecture, the richest connections are made locally between logic clusters and only a small amount (if any) are made through longer wire segments. Examples of FPGAs with this type of routing structure include the Algotronix CAL FPGA [234] (which later became the Xilinx XC6200), the CLi/Atmel 6000 FPGAs [162], and the Plessey/Pilkington ERA [215]. For the most part, these architectures tend to be very fine grained (2 or 3 input functions), so logic clusters are really very simple and have a single logic element in them. To aid with routing, the logic cells themselves were designed so they could be used as a part of the routing network between other logic elements. The cellular style of routing has fallen out of favor for several reasons:

- the delays for combinational paths can be significant for circuits that require more than nearest neighbor routing;
- CAD tools seemed to have significant difficulty efficiently mapping and routing circuits to the architectures (though, this may be true, in part, due to a lack of investment in the tools); and
- the area and delay costs of fine-grained architectures and routing are significant due to the amount of programmable logic and routing required to implement a function.

The final point can be mitigated somewhat if the design can be heavily pipelined since each cell has a flip-flop, but few designs come maximally pipelined. Recently, Cell Matrix [131], an FPGA-like architecture proposed for fabrication with nano-scale technologies, has revived interest in this routing architecture partially because of its relative simplicity to fabricate due to regularity.

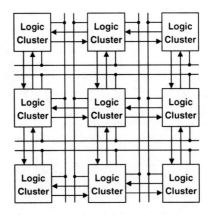

Fig. 2.9. Cellular Routing Architecture

Finally, for completeness, the row architecture for routing is illustrated in Figure 2.10. This routing architecture is mainly found in one-time programmable FPGAs such as many of Actel's anti-fuse FPGAs and is, therefore, not commonly seen in FPGAs used for reconfigurable computing. The architecture chiefly uses horizontal interconnect channels to route signals between two logic clusters. As suggested in the figure, Actel FPGAs (and others) do generally provide some vertical routing despite the bias toward row-based routing channels. For instance, in the Act-1 and later FPGAs, Actel used vertical wires to route the outputs of the logic clusters to several adjacent routing channels and provided some long wires that spanned even more row routing channels. Though not illustrated in the figure, row-based routing architectures generally used segmented wires within the routing channels to reduce routing delays for short paths.

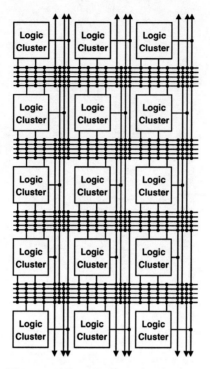

Fig. 2.10. Row Routing Architecture

Programmable I/O Architectures

Unlike logic and routing architectures, the basic input/output (I/O) architecture, as shown in Figure 2.11, is reasonably similar across FPGA families. In effect, the I/O blocks have tri-state buffers for the outputs and input buffers for the inputs. The tri-state enable signal, the output signal, and the input signals can be individually registered within the I/O block or can be left unregistered based on how the I/O block is programmed.

In modern FPGAs, a wide variety of additional features can be found that significantly enhance and complicate this basic structure. For instance, the Xilinx Virtex-4 architecture's I/O blocks [427] provide the following features, among others:

- more than 50 variations of I/O signaling standards some of which include features such as:

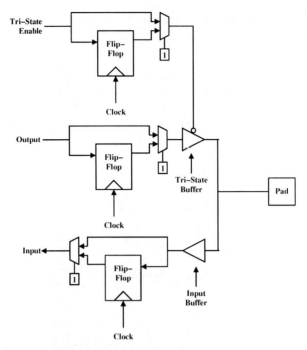

Fig. 2.11. I/O Block Architecture

> – digitally controlled impedance to eliminate the need for terminating resistors to deal with transmission line effects and
> – differential signaling to improve signal integrity;

- double-data rate registering with multiple modes for presenting and receiving data to and from the internal logic; and
- programmable input delays.

2.1.2 Specialized Function Blocks

Over time, the basic FPGA architectures that we have described above have been further developed through the addition of more specialized programmable function blocks. These blocks—such as embedded memory, arithmetic logic, high-speed serial I/O, and even embedded microprocessors—have been added due to a frequent need for such resources within FPGA systems and applications. The result is that many recent FPGAs are a more heterogeneous mixture of resources than early FPGAs. In the next few paragraphs, we will briefly describe the resources that have been made available in recent FPGAs.

Embedded Memory

Memory, of course, is a basic component of most digital systems and, though flip-flops can be used for memory, they are very inefficient for creating memories of any depth. Starting with the XC4000 series of FPGAs, Xilinx made the LUTs used for logic flexible enough to be used as asynchronous 16x1 RAMs in user designs. In later architectures, Xilinx has designed the LUTs so that they can operate as synchronous RAMs, dual-ported RAMs, and shift registers (1 to 16 stages). Logic also exists in the logic clusters to compose these smaller RAMs into wider or deeper RAMs. Being able to use LUTs for RAM is a feature unique to Xilinx FPGAs.

Though LUT-based memory is better than using flip-flops when implementing deeper memories, it was not long until most FPGA vendors started to include more dense blocks of SRAM within the architectures, with Altera leading the way with its FLEX 10K FPGAs [13]. Today, Altera, Xilinx, Actel, Atmel, Quicklogic, and other FPGA vendors include these larger SRAMs that have from hundreds to thousands of bits. Most of the RAMs tend to be on the range of 1- to 4-Kilobits (Kb), though, in the Stratix II architecture, Altera has three granularities of RAMs (576 Kb, 4.5 Kb, and .56 Kb) available to the designer. In many cases, the RAMs aspect ratio can be programmed. For instance, a Xilinx Virtex 4-Kb RAM can operate in the following possible modes: 4096x1, 2048x2, 1024x4, 512x8, 256x16 (where the aspect ratios are given as *depth* x *width* in bits). Besides having programmable aspect ratios, some FPGAs' embedded RAMs can operate as content-addressable memories (CAMs) [14], dual-ported RAMs, and/or FIFOs.

Though the total memory available on chip may only be as much as 1 MB for the largest FPGAs, one of the key benefits of these on-chip memories is the large number of memory ports available and the aggregate memory bandwidth that is possible, providing a significant advantage to very parallel applications that require significant memory bandwidth. For instance, the largest announced Stratix II (EP2S180) theoretically can provide a maximum aggregate memory bandwidth over 30 Gb/s through the 3414 ports of its 1707 RAMs (assuming each memory operates dual-ported and operates at maximum frequency) [15].

Embedded Arithmetic Logic

Beyond the basic carry logic and even adders provided in the logic elements and logic clusters, many FPGAs have started to include 18x18 multipliers or so-called "Digital Signal Processing" (DSP) blocks as separate, additional resources. In general, the DSP blocks provide addition/subtraction, multiplication, and multiply-accumulate (MAC) operations with a high degree of configurability. The MAC operations are useful in finite-impulse response (FIR) filtering, a common DSP operation (see Chapter 5 for a discussion on FIR filtering and FPGAs). A detailed description of these blocks for the latest

FPGAs from Xilinx and Altera is out of the scope of this chapter, especially, since the DSP blocks can be so flexible and complex (see [16, 429] for more detailed information), but Table 2.2 provides a summary of some of their features.

Features	Xilinx Virtex-4 XtremeDSP	Altera Stratix II DSP Block
Multiply-accumulate	48-bit	52-bit
Multiply	18x18, multi-precision (iterative)	8 9x9, 4 18x18, 1 36x36
Add/Subtract	Yes	Yes
Complex multiply support	Yes	Yes
Integrated FIR support	Yes	Yes
Saturating arithmetic	No	Yes
Routing arithmetic	Yes	Yes
Fixed point representation	No	Yes
Barrel Shifter	Yes	No
Iterative functions	Extended width multiply, Integer division, Integer square root	Possible(?)

Table 2.2. Xilinx Virtex-4 and Altera Stratix II DSP Block Features

High-Speed Serial I/O

Considering that many FPGAs are used in high-throughput telecommunications equipment, the recent addition of multi-gigabit serial transceivers (MGSTs) as I/O blocks may not be too surprising. These blocks perform full-duplex serialization and deserialization functions (SERDES), provide encoding/decoding functions (for instance, 8B/10B), and some error control logic. The first production FPGA to have this capability was Xilinx's Virtex-II Pro (up to 6.25 Gb/s full-duplex per channel). Altera's Stratix GX family (up to 6.25 Gb/s full-duplex per channel) and Xilinx's Virtex-4 FX family (up to 11.1 Gb/s full-duplex per channel) of FPGAs also provide this capability. Taking the concept of high speed communication even further, the Virtex-4 FX family of FPGAs has two or four dedicated 10/100/1000 Mb/s Ethernet Media Access Controllers (MACs) to provide designers with a more complete solution for Ethernet serial I/O solutions.

Embedded Microprocessors

Lastly, FPGA manufacturers have integrated full, dedicated microprocessors with their FPGA logic to perform low-bandwidth and/or control-intensive functions such as implementing TCP/IP stacks. With this capability, complete (or near complete) embedded systems can be implemented with a single device. The first commercial FPGA to integrate a microprocessor with its logic fabric was the Altera Excalibur. When announced the Excalibur was the integration of either an ARM- or MIPS-based 32-bit RISC processor core with an APEX-20KE-like logic fabric. Currently, only the ARM-based solution is available from Altera. As illustrated in Figure 2.12, the processor has dedicated external memory interfaces, a UART interface, a programmable interrupt controller, and other resources. Two dedicated AMBA high-performance buses (AHBs) exist in the system, providing the processor access to two different tiers of devices. The processor communicates with the programmable logic through the secondary AHB or through some dual-port SRAM that connects to both buses.

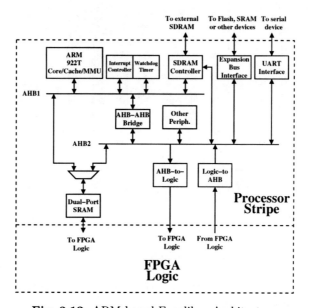

Fig. 2.12. ARM-based Excalibur Architecture

Later, with the Virtex-II Pro and Virtex-4 FPGAs, Xilinx produced FPGAs with integrated integer PowerPC microprocessors. One significant difference between the Xilinx approach to processor-logic integration and that of Altera is that the Xilinx approach places the microprocessor as an island within the FPGA logic, as illustrated in Figure 2.13, with interfaces to on-chip SRAM but no dedicated processor or peripheral buses. These buses must be implemented using FPGA logic, if they are desired. This provides the designer with the flexibility to define the architecture of the embedded system, but also means that the processor cannot perform useful work without configuring the FPGA logic, unlike the situation with the Excalibur. Note that the processors in Xilinx FPGAs replace any logic functions that would have normally occupied that area of the chip, but some of the routing architecture is maintained despite the presence of the PowerPC.

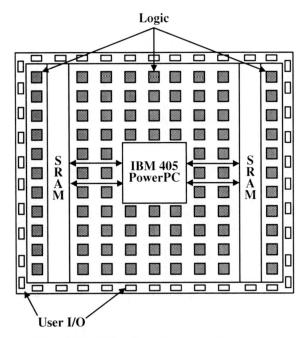

Fig. 2.13. Xilinx-Style Processor Integration

2.1.3 Programming Architecture

Reprogrammable FPGAs generally use SRAM to store the configuration data. The programming architecture of SRAM-based FPGAs is an important factor in the use of these FPGAs for reconfigurable computing. By programming architecture, we mean the way in which the FPGA's configuration data is structured and how it is provided to the FPGA to program its resources.

Many approaches to programming architecture have been developed for FPGAs, but the main characteristics of interest include the programming bandwidth, the granularity of the accesses, on-line programmability, and the ability to read out the programming data. We will discuss each briefly and describe their impact on reconfigurable computing.

Programming bandwidth, as you might guess, is the rate at which configuration data can be sent to the FPGA. When fast reconfiguration of an FPGA is required—such as when an application is time multiplexing logic on the hardware, the programming bandwidth is one factor that determines how fast the FPGA can be reprogrammed. The width of the programming interfaces are generally either serial or 8-bit interfaces and can run at tens of MHz. The Altera Stratix II FPGA, for example, can obtain a bandwidth of about 800 Mb/s for configuration in the Fast Passive Parallel mode—an 8-bit interface running at 100 MHz. As another example, the Xilinx Virtex-II Pro can operate up to 400 Mb/s. A common serial interface is the Joint Test Action Group's boundary scan interface (IEEE 1149.1)—which is often referred to as simply JTAG—that has been adapted for configuration. With JTAG, Altera's Stratix II FPGA achieves only about 10 Mb/s, where many Xilinx devices can have a bandwidth of up to 33 Mb/s. Some FPGAs have proprietary serial programming interfaces as well with higher data rates.

The granularity of the FPGA's programmability determines how many resources are configured with the smallest block of programming data. As far as the FPGA chip designer is concerned, this granularity has a significant effect on the cost of the internal logic used to perform the programming. Like with other forms of memory, the more addressability that is required within the programming data, the more logic that is needed to provide that addressability.

With respect to reconfigurable computing, the main interest in the granularity of the FPGA programming data is related to another aspect of an FPGA's programming architecture—the support provided by some FPGAs to perform on-line programming (or dynamic configuration) of a portion of their logic. This allows the FPGA user to modify the circuit as it executes. The programming granularity, then, has an impact on the smallest amount of data and logic that can and must be affected when making small changes to a circuit (for instance, reloading the ROMs holding coefficients used in some arithmetic or DSP algorithm).

As an example, the Xilinx Virtex series of FPGAs (Virtex, Virtex-E, Virtex-II, Virtex-II Pro, Virtex-4) allow partial reconfiguration of the FPGA. The smallest amount of data that can be programmed is called a frame, which is typically hundreds to thousands of bits. So, to partially reconfigure the FPGA logic at run time (sometimes referred to as run-time reconfiguration [417]), an entire frame of configuration data must be rewritten to the FPGA for the smallest change. By contrast, the now unavailable Xilinx XC6200 [81] (a later commercial version of the Algotronix CAL FPGAs [234]) allowed the user to send as few as 8, 16, or 32 bits per configuration operation and a mask could be used to restrict the modifications of a programming data write to a specific bit or set of bits.

The ability to read an FPGA's programming data and other information through the configuration interface—a capability sometimes referred to as "readback"—has also had an impact on how FPGAs are used in reconfigurable computing. First, it can be used to ensure that the programming data stored in the FPGA is correct—a concern when using FPGAs in space [161] or even in large quantities on the ground [151]. Next, it can be used as a way of sampling the state of a user's design, as with the Xilinx Virtex series of FPGAs as well as other such as the Xilinx XC6200. For instance, with the Xilinx Virtex FPGA, the outputs of the slices and the outputs of the IOBs (to the outside world and to the rest of the FPGA) can be sampled and read out through the FPGA's configuration interface along with the contents of the various RAMs. This can be used as a means of communication between an FPGA design and an external host (e.g., [63]) and it has been used for helping designers debug reconfigurable computing applications [181, 182, 205].

In addition to above characteristics, several modern FPGAs include programming data compression and programming data encryption capabilities. Compression, of course, reduces the amount of data that must be sent to the FPGA to program it, thus, reducing the storage requirements of the programming data and reducing programming time. The encryption capabilities have been added to FPGAs to prevent people from stealing configuration bitstreams from an FPGA system as it is being sent to the FPGA, say over the Internet or on the same system board. Though people can read the encrypted configuration data freely, a decryption key is stored on the FPGA (and kept available through battery power) before sending the system to the customer so the FPGA can decrypt the bitstream without allowing others to easily steal it. Readback of the configuration data is disabled when the configuration bitstream is encrypted to prevent unauthorized access to the actual programming data. Bitstream encryption is one of several components necessary for secure programming of reconfigurable computing machines through networks such as the Internet.

As a final note on programming architecture, several research FPGA devices [109, 265, 355, 396] have extended configuration in a novel way. These FPGAs support switching among multiple configuration states stored on the FPGA . This allows the FPGA circuit to change rapidly while the data in the FPGA either remains in place or is itself swapped in and out of the array, thus, emulating a much larger FPGA. Once the multiple, on-chip configuration memories are loaded, this considerably reduces the external programming bandwidth required to context switch among FPGA configurations since switching between device contexts (i.e., individual device configurations) is all done through accesses to on-chip configuration memory.

2.2 Coarse-Grained Reconfigurable Arrays

The configurability of fine-grained reconfigurable logic devices allows designers to specialize their hardware down to the bit level, meaning that, if an application requires 7- or 17-bit arithmetic for an operation, the hardware can directly accommodate what is needed. The configurability makes devices, like FPGAs, suitable to implement a large variety of functions directly. This configurability comes at a significant cost in terms of circuit area, power, and speed. Every level of configurability requires more multiplexing, buffering, routing, and/or memory, thus, requiring more transistors and their interconnection.

In recognizing these costs, many researchers [11,51,69,72,99,174,198,201, 256,280,291,294,380,408,439] have studied how to use arrays of more coarse-grained reconfigurable operators as the basis for reconfigurable computing machines. Besides the potential advantages in terms of circuit area, power, and speed, many researchers have pursued the use of coarse-grained reconfigurable arrays (CGRAs) with the hope that they would be easier targets for higher level development tools. However, as pointed out in Hartenstein's brief overview of 19 different coarse-grained architectures [197], the challenge with using CGRAs is that there is no universal form that is effective for all applications. To be more efficient, CGRAs are optimized in terms of operators and routing for some specific problem domain or set of domains. Thus, an application must be well matched with the array to realize dramatic improvements over other reconfigurable solutions. For example, an application that performs many 8-bit operations wastes a significant amount of logic if the CGRA uses 32-bit ALUs and operators. Considering the large number of CGRAs, we will briefly describe just a few notable research architectures as well as a few recent commercial architectures to illustrate some interesting past and current examples of CGRAs.

2.2.1 Raw

The first coarse-grained architecture we will briefly describe is probably one of the most coarse-grained—the Raw chip [386,408] from MIT. Effectively, it is a two-dimensional array of programmable tiles, each having: a 32-bit MIPS-like microprocessor, local instruction and data caches, and a 32-bit pipelined floating point unit (FPU) as well as several routers and wiring channels to support the four on-chip 2-D mesh networks. At one point, the designers had considered putting reconfigurable logic into each of the tiles, but this option was later abandoned. Figure 2.14 roughly illustrates a 4x4 Raw array—the array size that has actually been fabricated [386].

Unlike a traditional bus-based multiprocessor system, the Raw machine uses a switched network for communicating directly between the processors. The length of wires in the architecture are bounded by the width of a tile and each routing segment is registered on tile boundaries. Thus, to ensure that the array can scale, the width and height of tiles are limited by the distance a signal can travel in a single clock cycle. Two of the mesh networks are statically scheduled for predictable performance while the other two allow for dynamic scheduling through wormhole routing. Programmable routers handle the flow of data through the system and FIFOs help synchronize the transfer of data between tiles. The statically scheduled networks provide the best performance and significantly lower latency and lower overhead than traditional multiprocessor communication, enabling very fine grained coordination of operations between processors. For instance, an operand produced by one tile can be processed the following cycle by its neighbor, enabling the RAW machine to operate as a pipeline of ALUs or FPUs for a stream of data and not just

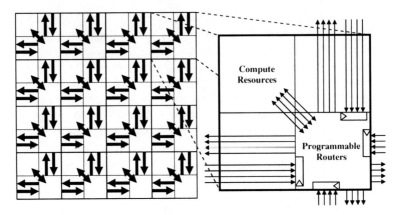

Fig. 2.14. Raw Microprocessor Array Architecture

a traditional multiprocessor system on a chip. The dynamic networks, on the other hand, are used for less predictable or bursty forms of data movement, such as cache misses, some forms of data I/O, and operations that happen only occasionally.

Also unlike a more traditional multiprocessor on a chip where hardware handles many issues, the Raw device requires that much of its internal operation be handled by the compiler. For instance, cache coherency, cache misses, and the routing of data over the internal wiring must be handled by the compiler (or user).

2.2.2 PipeRench

Another example of a novel CGRA is PipeRench [174, 356], a project from Carnegie Mellon University. The chief goal of PipeRench was to develop an architecture that effectively employs run-time reconfiguration for hardware virtualization. Thus, an application that does not physically fit on a particular PipeRench's available resources can still execute. Using a coarse-grained architecture in this case greatly reduces the amount of configuration data that must be quickly swapped in and out of hardware regions.

Figure 2.15 illustrates the general architecture of this CGRA. The hardware is organized in pipeline stages called "stripes". Each stripe consists of 16 processing elements (or PEs) that contain 8-bit-wide logic and an 8-entry register file. The PEs within a stripe are all interconnected, reducing placement and routing issues. Note that the physical stripes are cleverly interleaved with the appropriate register file interconnections to create a ring structure out of the stripes—something that is important for virtualization. The on-chip logic for controlling the configuration and the virtualization functions is not shown in the figure.

With regards to PE logic, each PE contains shifters and multiplexers that can be configured to operate on inputs and the PE's main functional unit

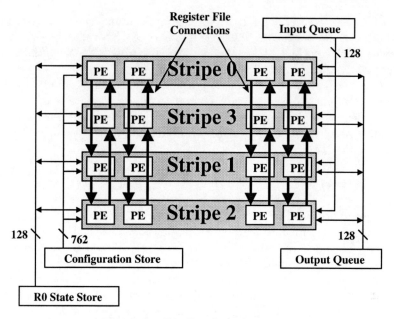

Fig. 2.15. PipeRench Architecture

is a collection of 8 3-bit LUTs, one for each bit of the operand width and each configured with the same function. In addition to the LUTs, specialized carry logic is also included to support fast addition. Due to the interconnect available, the PEs can be easily combined to form wider operations, including wide shifts using the input shifters. Only 42 bits are required to configure a PE while 672 bits are needed for an entire stripe.

As mentioned above, the register file provides 8 entries, but it also is a key part of the communication of data within the architecture. A PE's output can be written to any of the 8 registers and those registers not written to by the PE are overwritten by the previous stripe's values for those registers— basically, implementing the pipelining of operands. Further, Register 0 (R0) plays a special role. If a stripe operates as the first stage of a virtual pipeline, its R0 registers are loaded by the global input bus to provide the data for the pipeline. If a stripe operates as the last stage of a virtual pipeline, then the R0 registers' outputs are written to the global output bus.

Regarding virtualization, if a virtual stripe must be swapped out of a physical stripe to make room for more logic, the R0 registers for the stripe are stored away in the R0 State Store before reconfiguring the stripe. Likewise, when a virtual stripe is restored into a physical stripe, its configuration and R0 state are restored from the Configuration and R0 State Stores, respectively. One pass through the virtual pipeline is made for each set of new inputs to the first virtual pipeline stripe. With its interconnection architecture and this state configuration/restoration support, the PipeRench architecture is thus

constructed to easily support PipeRench applications across a wide range of
PipeRench devices having different numbers of physical stripes. The actual
device fabricated by CMU was a 16 physical stripe device with the ability to
store 256 virtual stripes [356].

Considering the communication channels mentioned above, PipeRench
supports applications with only limited feedback. Despite this, many data-
path-oriented applications—such as FFTs, DCTs, and many encryption algorithms—require only feed-forward structures and map reasonably well to the
architecture.

2.2.3 RaPiD

Another important linear CGRA architecture called RaPiD [99–101, 134] (for
Reconfigurable Pipelined Datapath) was developed by the University of Washington. The goal of the architecture were to develop a high-performance
coarse-grained reconfigurable device that could be targeted to specific application domains and could support application development at a relatively
high level (i.e., not doing hardware design as with FPGAs). To help with the
second of these two goals, RaPiD's architecture and application development
system were co-developed to ease application development for the end user.
The RaPiD architecture and related domain-specific reconfigurable architectures continues as a topic of research even today [86, 327].

As illustrated in Figure 2.16, this architecture is structured so that data
is streamed through a mix of different coarse-grained function units having
a common data width (generally between 8- to 32-bits wide) and the interconnection between these function units is through a single, flexible routing channel. A "Streams Manager" provides the architecture with streams of
data from external memory or other sources and likewise receives the output streams from the CGRA and writes the data to external devices. During
operation, the data pipeline does not need to be statically configured—the
RaPiD architecture allows some resources' configurations (e.g., input muxes,
ALU functions, etc.) to be altered during the operation of the application
so the data flow can be somewhat dynamic, changing based on the application's needs. The configuration of the dynamically changeable resources is
controlled cycle by cycle using the "Instruction Generator" and "Configurable
Instruction Decoder."

As for the function units, the mix and number of function units is chosen
based on the application domain for which the device will be used. The mix
of function units is not necessarily limited to ALUs, multipliers, registers, and
RAM, but can be any function that can be well utilized in a given application
domain. For instance, a RaPiD for the communications application domain
might have a Viterbi decoder as a function unit. A configurable delay element
providing a zero to three-stage delay exists at the output of each unit to help
with the scheduling of operations within the pipeline.

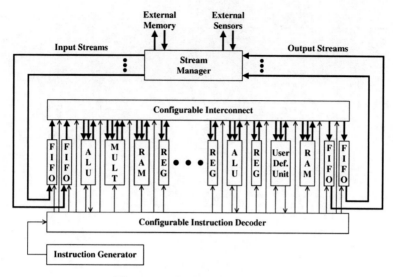

Fig. 2.16. RaPiD Architecture

The routing channel has segments of various lengths to support communications at different distances between function units. Further, as mentioned above, the interconnection among modules can be dynamic such as through selecting different inputs using the function units' input multiplexers. Unlike the other coarse-grained architectures we have described, some of the routing can be driven in one of two directions, allowing for feedback and flexibility in the mapping of applications. The direction a routing segment is driven, though, cannot be dynamically changed.

The "Stream Manager", which produces the input data and consumes the output data, is essentially a memory interface with an address generator and FIFO for each input or output stream (the FIFOs are explicitly illustrated in Figure 2.16). The flow of data through the architecture and instruction generation (or sequencing) are decoupled. A RaPiD array does provide a synchronization mechanism, though: the RaPiD array is halted upon the read of an empty input FIFO or the write of a full output FIFO.

2.2.4 PACT XPP

As an example of a commercial CGRA, the eXtreme Processing Platform (XPP) [36,317] is a computing array with a data-driven processing model and hierarchical configuration management developed by PACT Informationstechnologie GmbH. Like many CGRA architectures, the PACT XPP was developed to handle streaming data applications such as signal or media processing. As a product, PACT provides the CGRA as intellectual property to their customers for developing custom VLSI designs—a common model for commercial CGRA products.

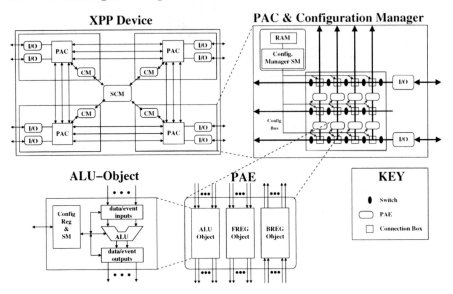

Fig. 2.17. PACT's eXtreme Processing Platform

Figure 2.17 illustrates the architecture at four different levels. The top level is the XPP device, which consists of several Processing Array Clusters (PACs) with their associated configuration management hardware. The configuration management (CM) hardware includes a state-machine-based controller and local RAM. At the device level, a supervising CM (or SCM) controls the overall configuration of the device and this SCM can yet be controlled by SCMs external to the device, creating a configuration management hierarchy.

At the next-level down, the PAC itself is an array of Processing Array Elements (PAEs), connection boxes for routing between vertical and horizontal busses, and switches for segmenting the horizontal busses. Each PAC also includes I/O resources as well.

At the third level, each PAE contains three objects: a function object, a forward register (FREG), and a backward register (BREG). In Figure 2.17, an ALU Object is the function object, but other objects, such as RAMs are possible. The FREG and BREG objects provide vertical routing support as well as data control flow (FREG), counters (FREG), adders/subtracters (BREG), and barrel shifters (BREG).

At the lowest level, the ALU object can consume and produce data and event packets that are key to the data-driven computation model used by the architecture. Data packets, of course, contain the results of an operation while the event packets contain the condition or state bits resulting from the operation. The width of the data packets are the width of the basic architecture (e.g., 24 or 32 bits) while the width of event packets is only a few bits. Note that the configuration portion of the architecture can be affected by ALU operations and the event packets, allowing for event-driven initiation of configuration or reconfiguration.

The data-driven computation model is used in the XPP to ensure ease of application development—instead of having to worry about the exact timing of operations, the operators only process when all of their input packets are available and their last output packet has been consumed by the next PAE. This data-driven model has also been augmented with additional handshaking between PAEs and CMs so that a CM knows when a PAE can be reconfigured and when it is busy with a computation.

The hierarchical nature of configuration management has several effects. First, it provides a way for scaling the configuration of large systems or devices. Next, its allows configuration to be performed independently in the different chip regions. A corresponding result is that different portions of the XPP can be executing unrelated functions. Finally, the self-configuration capability can also be used either locally or globally throughout the system.

2.2.5 MathStar

Another commercial example of a CGRA architecture is the Field-Programmable Object Array (FPOA) produced by MathStar [203]. Like RaPiD, the MathStar architecture is intended to be optimized for a particular application domain and customers can specify the needed mixes of function units (called Silicon Objects) to meet their application needs. To make this a cost effective and fast time-to-market proposition, MathStar's FPOA is structured so that any function unit type can fit at any position in the device's two-dimensional object array. The result of this engineering is that an FPOA with a custom function unit mix can be ready for fabrication in less than a month with 1-GHz internal operation speeds without having to solve any additional analog signaling issues since the Silicon Objects are pre-engineered to deal with such issues.

Figure 2.18 is a conceptual illustration of the architecture. As mentioned before, the array can be heterogeneous or homogeneous, depending on the particular mix of Silicon Objects chosen for the array. The array given in the figure is a mix of register files, ALUs, and multiply/accumulate units. Note that the architecture also supports various I/O standards (including high-speed serial I/O) as well as internal RAM. The available Silicon Objects also include a logic block consisting of four four-input LUTs, a CRC generator, content addressable memories, and external memory interfaces.

As for the routing architecture, the FPOA supports 21-bit busses to communicate 16 data bits, a one-bit data-valid flag, and 4 control/state bits. Generally, the data and data-valid bits are handled as a unit while the control bits can be configured independently. Depending on the Silicon Object, the control/state bits can provide such information as the sign of the data, a carry bit, or a start-of-packet marker or they can be used to control the function of

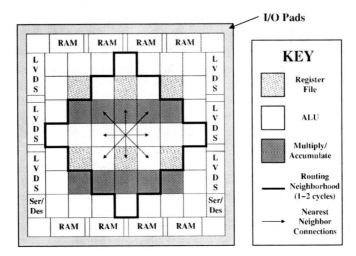

Fig. 2.18. MathStar's FPOA Architecture

the Silicon Objects, such as selecting the function of the ALU or the modes for the multiply-accumulate unit.

As the arrows in Figure 2.18 illustrate, each Silicon Object can communicate directly with its 8 immediate neighbors with, at most, four unique values. The values are available within a single clock cycle. The figure also illustrates that, with one level of pipelining, a Silicon Object can route its signal to any of 24 other cells within its extended neighborhood. Using more levels of pipelining, a Silicon Object's output signals can reach the rest of the array.

2.3 Summary

In this chapter, we have discussed the internal structure of FPGAs. Starting from simple homogeneous tiled arrays of logic blocks, I/O blocks, and interconnect, FPGAs have become complex systems on a chip, with elaborate logic clusters, a rich memory hierarchy, dedicated arithmetic function units, and high-speed serial I/O. The capacity of FPGAs has also greatly increased, which makes possible larger reconfigurable computing applications. The ability to reprogram SRAM-based FPGAs, either entirely or partially, is also an important feature for reconfigurable computing. Finally, we discussed how new coarse grained architectures have been developed that trade off FPGA flexibility for increased performance and lower power within specific application domains.

3

Reconfigurable Computing Systems

In this chapter, we will discuss general purpose computing systems that incorporate FPGAs into the system architecture. While modern FPGAs include processors, memory blocks, and built-in I/O interfaces on-chip, reconfigurable systems, even those with a single FPGA or tiled processor array contain off-chip memory and I/O resources as well. Since reconfigurable computing is concerned with parallel operations at any level of granularity, we will motivate the roles that FPGAs can play by first discussing parallel processing models and how they might use reconfigurable logic. We will then survey the field of reconfigurable processing systems.

3.1 Parallel Processing on Reconfigurable Computers

Reconfigurable computing systems derive high performance by exploiting parallelism at multiple levels of granularity, from instruction through task level parallelism. In this section we introduce the levels of parallelism and discuss the use of reconfigurable hardware at various granularity of parallelization.

3.1.1 Instruction Level Parallelism

The lowest level of granularity we consider is instruction-level parallelism. In conventional microprocessors, instruction-level parallelism is exploited in the micro-architecture of a superscalar processor. By having multiple instructions in progress in different stages of completion, the superscalar processor is able to complete more than one instruction in a clock cycle.

Very Long Instruction Word (VLIW) processors offer another method for fine-grained parallel operation. A VLIW processor contains multiple function units operating in parallel. In Figure 3.1, the instruction word contains fields for two integer operations, two floating point operations, two memory operations, and a branch. To compile for a superscalar processor, the compiler

simply generates a sequential instruction stream, and the processor parallelizes the instruction stream at run time. In contrast, the VLIW processor
executes the instruction word generated by the compiler, requiring the compiler to schedule concurrent operations at compile time.

Co-processor parallelism is achieved within a single instruction stream.
A customized parallel instruction is performed by co-processor. Examples of
co-processors include MMX/SSE units or vector units. Instructions for the
co-processor are integrated into the instruction set of the processor. The co-
processor shares register files and other internal state with other arithmetic
units, such as the floating point units, as shown in Figure 3.2.

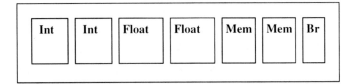

Fig. 3.1. A VLIW Instruction Word

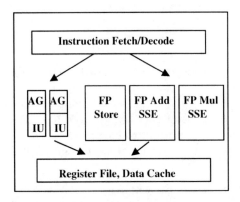

Fig. 3.2. SSE co-processor

Instruction level parallelism is of central importance to RC systems as
well. In contrast to conventional processors in which the instruction unit is
optimized to a very large class of applications, it is possible (and necessary)
in RC systems to tailor instruction level parallelism uniquely to each specific
application. In a VLIW processor the instruction format is determined by
the number and type of parallel function units, which are are fixed when
the processor is designed. On an FPGA these units are constructed from
the configurable logic blocks. Their number, width and type are arbitrary,

and can be optimized to the application at compile time. If partial dynamic reconfiguration is available, the function units' design may even be modified at run time. Thus an "instruction" as interpreted by the reconfigurable system is an arbitrary collection of related logic circuits, in which the number and type of arithmetic units is optimized to each application.

Figure 3.3 illustrates the data flow within an RC instruction composed of 11 levels of operation. At each level, multiple operations occur in parallel. For example, the first level contains six multipliers. If the instruction itself is pipelined, at every clock cycle, there are 23 arithmetic operation occurring in parallel. The specific application being mapped to reconfigurable logic determines the widths and data types of the operations. Data dependencies within the application determine to what extent the customized instruction can be pipelined. An instruction for a reconfigurable computer is analogous to an entire subroutine on a conventional processor.

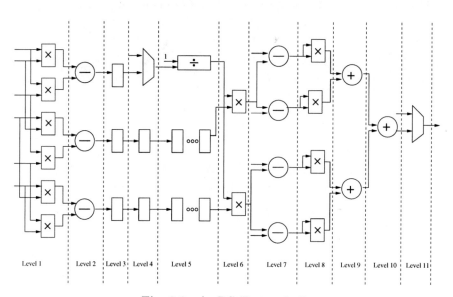

Fig. 3.3. An RC "Instruction"

3.1.2 Task Level Parallelism

Task level parallelism has two major categories. The relationship between tasks can be either peer-to-peer or client/server. In a peer-to-peer parallel system the parallel activity can be at a process level or the finer granularity thread level. In process level peer-to-peer systems, illustrated in Figure 3.4, each process has its own separate address space. In order to communicate state, a process must send a message and the destination process(es) must explicitly receive the message. Several different sorts of messaging protocols may be used. Messages may be buffered, asynchronous, and high latency, as with Message Passing Interface (MPI). They may use a low latency streaming protocol, with queues, FIFO buffers, or "valid bit" semi-synchronous communication. Finally, the messages may be tightly synchronized streams, in which delays are determined at compile time and are pre-compiled into clock-synchronized processes. The latter protocol is typical of DSP algorithms.

At the thread level of peer-to-peer parallel processing (see Figure 3.5), the threads share an address space, and can communicate through shared memory or messaging. Signaling and synchronization mechanisms for thread-based processing include critical sections and mutual exclusion, or barriers.

In a client/server parallel model, the client can request service from server, or the client can request work from server. In the former (shown on the left in Figure 3.6), the compute resource is centralized in the server, whereas in the latter, shown on the right, the clients perform compute intensive tasks and the server merely serves as repository for the task description.

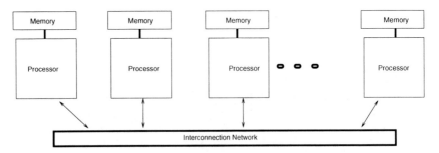

Fig. 3.4. Process Level Parallelism

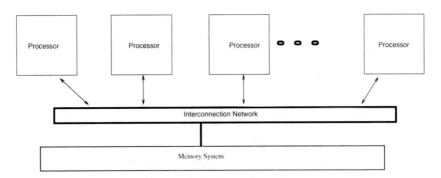

Fig. 3.5. Thread Level Parallelism

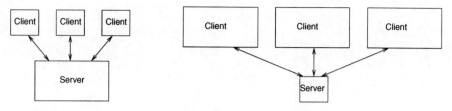

Fig. 3.6. Client Server Model

Reconfigurable logic can play a role in these large granularity parallel activities as well. In peer-to-peer processing, hardware processes communicate with software processes. This scenario is particularly common in embedded processing, in which hardware can handle data acquisition and front end processing, while software performs data analysis and decision making. The most common forms of hardware-to-software communications are through memory buffers or with stream communication, which is illustrated in Figure 3.7. In this example, two hardware processes (Processes 1 and 2) communicate via a hardware-to-hardware stream, Process 2 and 3 communicate via a hardware-to-software stream, and Processes 3 and 4 communicate via software-to-hardware stream. RC systems tend more toward static, application-specific communication channels rather than dynamic streams as proposed in dedicated stream processors such as Imagine [6]. This is because statically determined stream communication paths are orders of magnitude more efficient when mapped to FPGAs.

RC plays an important role in client/server processing, too. RC nodes may perform data/compute tasks, for example for cryptography, bio-informatics or network security applications. RC nodes may also serve as data acquisition nodes, as shown in Figure 3.8. In this scenario, multiple data sources feed into reconfigurable logic nodes that pre-process the data and supply it to other compute nodes on a network.

3.2 A Survey of Reconfigurable Computing Systems

Parallel processors based on FPGAs first appeared in the early 1990's and have continued to flourish over the succeeding decades. Reconfigurable computing systems can be grouped into four categories:

- I/O Bus Accelerator
- Massively parallel FPGA array
- Reconfigurable Supercomputer
- Reconfigurable Logic Co-processor

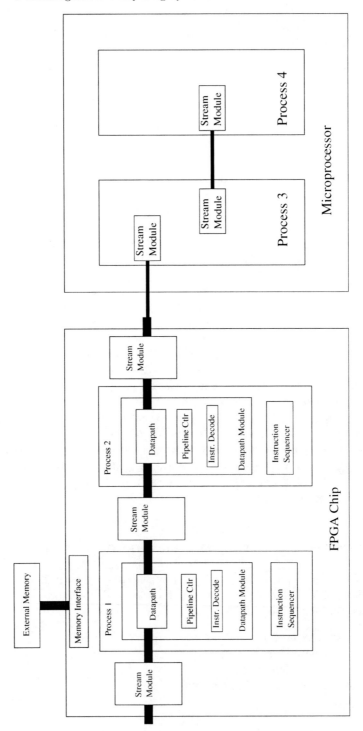

Fig. 3.7. Stream Communication

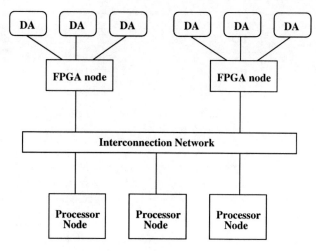

Fig. 3.8. Data Acquisition

3.2.1 I/O Bus Accelerator

The earliest reconfigurable computer to emerge in the early '90's was the accelerator Printed Circuit Board (PCB). In this architecture (shown in Figure 3.9), the I/O board contains FPGAs, inter-FPGA interconnect, on-board SRAM and/or DRAM modules, high speed (serial) interfaces to external devices, and an interface to the host computer's I/O bus. There are many variations possible on inter-FPGA interconnect. If there are few enough FPGAs and sufficient I/O pins on the FPGA chip, an all-to-all interconnect is possible. Otherwise, a ring interconnect may be used in which each FPGA is directly connected to its left and right neighbor. This was the interconnect used by the Splash 1 board ([168] with 32 Xilinx 3090 FPGAs. This board was built by the IDA Supercomputing Research Center in 1989. A programmable crossbar may be employed, allowing arbitrary interconnect among FPGAs, determined on a per-application basis. This arrangement was used in Splash 2 ([60]), a 1994-vintage board. A two-dimensional mesh interconnect was used by the PerLe 1 board built by the Digital Equipment Corporation Paris Research Lab ([47] in 1989. Modern commercially available FPGA I/O boards (for example, [305], [22], [128], [422], [403]) tend to contain a small number of large FPGAs. With only one or two FPGAs on the board, interconnect is no longer a major board-level architectural concern.

FPGA accelerator cards typically contain on-board buffer memory in the form of SRAM and/or DRAM. Unlike the narrow, deep cache memory hierarchy used on microprocessors, FPGA processors can best exploit very wide access to independent memory banks. In fact, it is not uncommon for FPGA boards to contain 5–10 SRAM banks, allowing, e.g., $64 * 10 = 640$ bits of concurrent memory access every clock cycle.

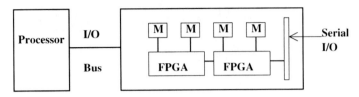

Fig. 3.9. Generic FPGA I/O Board

Since bandwidth between host processor and I/O card is high latency and low bandwidth, use of an on-board memory sub-system allows compute– and data– intensive tasks on the FPGAs to have direct, application-specific access to data arrays. One common design pattern ([108]) is for

1. the host to DMA data arrays to the board memory banks,
2. the algorithms on the FPGA to process data in those memories, perhaps even making several passes over the data,
3. the FPGA to write results results into memory, and
4. the host to read back the results via DMA.

Another design pattern is to double buffer the memories, so that the host can write input data to one set while the FPGA(s) process data from another set of memories. This approach is particularly effective with a board such as the USC/ISI Osiris PCI board (Figure 3.10) with 10 independent SRAM modules.

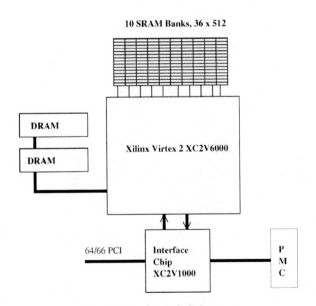

Fig. 3.10. Osiris RC System

In many cases, especially with embedded processors using the VME bus, I/O bandwidth between host and FPGA board is not high enough to keep up with the FPGAs' compute rate. Instead, data is presented to the FPGA processors directly from external devices such as A/D converters, network interfaces, or framegrabbers. The FPGA can then process the data, often with a dramatic reduction in bandwidth. The data can go to the host computer or directly back on the network as in Figure 3.8.

3.2.2 Massively Parallel FPGA array

A second FPGA-based architecture generalizes the FPGA accelerator board by putting together a very large collection of FPGA boards as a single computing resource, as shown in Figure 3.11. This architecture is particularly useful for logic emulation ([32], [37]) or very large scale computing [382]. The massively parallel FPGA array requires a high performance interconnect so that communication between FPGA boards can more closely match bandwidth between FPGAs on the same board. In the Virtual Wires project, the interconnect was a virtual resource. Each FPGA board communicated with its four nearest planar neighbors through eight bi-directional signals. To expand to multiple boards, each row or column of FPGAs on the periphery communicated to an adjacent board. To compensate for the lack of physical routing resource, a CAD tool automatically partitioned a large design across the FPGAs, time-multiplexing signals across the limited physical interconnect. This resulted in potentially a very slow logical clock (in the KHz) so that the same wires could be used for many connections occurring during the logical clock period.

In contrast, the Starbridge machine allocates FPGAs explicitly for routing and off-board communication. The Starbridge architecture has four to eight compute FPGAs on a board, with an additional three FPGAs devoted to communication. In contrast to this rich communications network on and between boards, the Starbridge uses PCI to communicate to a host computer.

Similarly, the Berkeley Emulation Engine 2 (BEE2) is a massively parallel FPGA array. Each board contains four compute FPGAs connected in a mesh with an additional control FPGA for off-mesh connection, a configuration similar to the Splash 2. Boards communicate over an Infiniband-based global communication tree. Every 16th board serves as a tree node in the interconnect tree.

3.2.3 Reconfigurable Supercomputer

In reconfigurable supercomputing, FPGA boards are included as acceleration nodes into a high performance cluster. The distinguishing factor in this architecture is that the FPGA board communicates with conventional processors as well as other FPGA boards through a high-bandwidth, low-latency interconnection network. Unlike the accelerator card architecture, communication

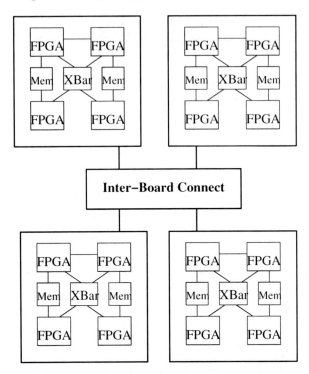

Fig. 3.11. Massively Parallel FPGA Array

with a host computer is an order of magnitude higher bandwidth, and latency is on the order of microseconds. This allows the FPGA computation to share state with microprocessor more easily, and thus the granularity of computation on the FPGAs can be smaller. In addition, there can be many FPGA boards on the interconnection network, allowing the aggregate system to solve very large supercomputing problems.

The first reconfigurable supercomputers of this form were developed by SRC Computers, Inc. The SRC architecture (Figure 3.12) exploits the DRAM interface of Pentium processors as a communication port to the "MAP" board containing high end Xilinx FPGAs. The SNAP ASIC manages the protocol and communication between a dual processor Pentium node and its associated MAP (FPGA) board. These microprocessor/FPGA units can be combined using commercial interconnect such as Gigabit Ethernet to form a cluster. For higher performance, a proprietary switching network connects multiple "MAPStations." In addition, MAP boards can communicate directly through chain ports, by-passing the interconnection network for large, multi-FPGA-board designs.

Another entry into the reconfigurable supercomputer arena is Cray Research, Inc., with its acquisition of Octigabay. The Cray XD1 combines AMD

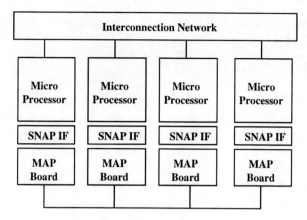

Fig. 3.12. SRC Reconfigurable Supercomputer

dual processor Opteron nodes, a proprietary "RapidArray" interconnection network, and FPGA boards. As with SRC, the interconnect network is low latency and high bandwidth. The Cray machine implements the network interface logic in an FPGA rather than a specialized ASIC.

Characteristics of the SRC and Cray machines are described in Figure 3.13.

Vendor	Proc	Comm BW	FPGA	Memory	Inter-FPGA interconnect
SRC	Pentium 4	4.8 GB/s	V26K	24MB	Chaining 4.8 GB/s
Cray	Opteron	3.2 GB/s	V2Pro50	32MB	RapidIO 3.125 Gb/s

Fig. 3.13. SRC and Cray Comparison

A similar approach has been followed by SGI, Inc. with their Reconfigurable Application-Specific Computing Platform. In the SGI architecture, the high performance microprocessor cluster consists of IBM POWER or Intel Itanium 2 processors interconnected through the SGI Non-Uniform Memory Access (NUMA) network. RASC extends the standard architecture by including special purpose accelerators such as FPGAs, which interface to the interconnection network through an ASIC, the "TIO." SGI reports maximum I/O bandwidth to the TIO as 3.2GB/s in each direction. In addition, the TIO provides direct, coherent memory access between FPGAs and system memory, a unique feature to the SGI architecture.

3.2.4 Reconfigurable Logic Co-processor

The reconfigurable logic co-processor is the ultimate coupling of traditional microprocessor with reconfigurable logic. As with parallel processors, it is only worthwhile to migrate functionality to hardware if the hardware execution time combined with the overhead of migrating to hardware is significantly less than the time to compute the function in software:

$$T_{hw} + T_{oh} << T_{sw} \qquad (3.1)$$

By allowing the processor and reconfigurable logic to share low level state, the T_{oh} component of the inequality can be greatly reduced, allowing for smaller granularity tasks to be profitably mapped to hardware. It also makes it feasible to dynamically reconfigure the logic during execution, analogous to context switching in conventional processors.

This notion of dramatically reducing latency and increasing bandwidth between microprocessor and reconfigurable logic by combining the two into a single entity has recurred since the early days of reconfigurable computing. In 1994, two processor-coupled architectures were proposed, the DPGA by De-Hon [107] and the PRISC [340], a RISC processor with programmable function units. Following on these early proposals, The OneChip design from Toronto ([418]) combined a MIPS-like basic function unit with reconfigurable function units, allowing both entities to share register files and memory. A single instruction stream controlled both standard and reconfigurable processors.

The Garp architecture [62] extended a MIPS core with a unique column-oriented reconfigurable logic array. The array situated between the processor and cache could communicate with the processor via four 32-bit data busses and one 32-bit address bus. Thus the processor's data cache was shared with the co-processor array. The architecture included a configuration cache to hold recently used configuration frames. The extended instruction set included commands to the co-processor. Simulation of the Garp processor predicted speed-up of 2–40 over an UltraSparc.

The Garp was the first project to develop a C compiler that automatically partitioned sequential code between software and hardware, automatically generated hardware for the custom logic array as well as interface logic to reconfigure the array dynamically during execution.

The first reconfigurable co-processor IC to be fabricated was the NAPA 1000 chip by National Semiconductor [350]. This chip used a 50 MHz 32-bit RISC core along with the CLAy reconfigurable logic array from National. The RISC processor, on-chip local memory, a reconfiguration controller, an interconnection network controller, and the reconfigurable logic array shared a common Core Bus (see Figure 3.14. The NAPA architecture was also the first to provide parallel processing of the NAPA 1000: the "toggle bus transceiver" was an interface to a high performance interconnection network for broadcast, data rotation, and data reflection communication patterns. The reconfigurable logic array could directly access local, dedicated scratch-pad memory. Computation on the logic array was optimized for a 32-bit linear, pipelined datapath. 32-bit columns could be dynamically reconfigured during operation.

Like Garp, the NAPA project included a C compiler, the NAPA C compiler [170]. In NAPA C, the programmer used pragmas to tag regions of code to be placed into hardware. This methodology is most effective when the application shows "90/10" behavior – 90% of execution time in 10% of the code. By placing such compute intensive blocks in hardware, the NAPA could achieve 1–2 orders of magnitude performance improvement over DSP processors.

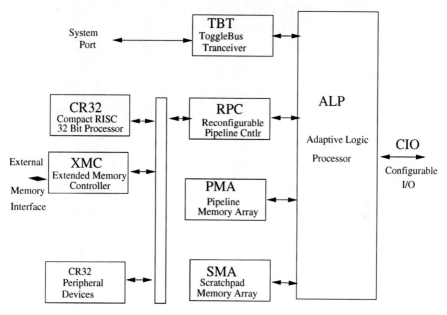

Fig. 3.14. NAPA 1000

Following these research proposals, commercial FPGA vendors began offering embedded processors in their FPGA product lines. The Triscend chip was one of the first such products, with an 8-bit microcontroller. More recently, Altera produced the Excalibur chip with an embedded ARM, and Xilinx followed with embedded PowerPC processors. Stretch, Inc. proposes a OneChip-like architecture in which the configurable logic shares register files with a Tensilica processor.

3.3 Summary

Reconfigurable logic has found its way into virtually every part of the parallel processing hierarchy. Millions of system gates are available for Instruction-level parallelism within the reconfigurable fabric. By sharing registers or internal busses with a conventional microprocessor, co-processor performance is even more accelerated through low latency, high bandwidth communication between processor and reconfigurable logic. In the I/O model, an FPGA array can process large granularity compute- and/or data-intensive tasks for the processor. Finally, by aggregating collections of FPGAs either as a standalone system or as part of a reconfigurable cluster supercomputer, parallelism can be exploited at a very large scale.

4
Languages and Compilation

As we have seen, reconfigurable computers are capable of parallelism at many levels from intra-operation parallelism (e.g., pipeline a × operation) to task level parallelism (e.g., communicating hardware tasks and software tasks). It would be ideal if automatic tools could profile, partition, parallelize, and compile existing code onto reconfigurable systems. It would be desirable to expression computation in a high level language and use the compiler to automatically detect compute-intensive portions and translate those into custom hardware instructions, leaving inherently sequential or infrequently executed code in software, automating the steps outlined in Figure 4.1. However, the reality is much different. As we will see in this chapter, research projects have attempted to realize this dream, but tools available today and in the near future require deep understanding of reconfigurable systems' strengths and weaknesses, considerable manual effort to partition the code, and familiarity with hardware design to map code segments to reconfigurable logic.

4.1 Design Cycle

In evaluating languages and compilers for RC systems, there are several criteria that should be considered:

- Abstraction Level. Is the abstraction level of the language algorithmic, behavioral, or structural? Is it textual or graphical? How is the system level described? How is the component level described?
- Automatic parallelization and partitioning. To what extent does the compiler provide automatic parallelization of sequentially expressed algorithms? To what extent does the compiler automatically partition computation between software and hardware?
- Loop level transformations? For algorithmic languages and compilers, does the compiler perform loop unrolling or pipelining? How do the loop level and instruction level pipelines interact?

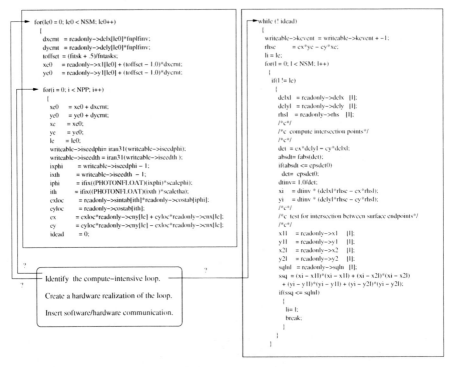

Fig. 4.1. Partitioning an RC Application

- Automatic instruction level parallelism (ILP) extraction. Can the compiler infer ILP and automatically schedule parallel activities?
- Memory hierarchy management. How are arrays and variable allocated to memory resources both on and off chip?
- Debug Support. What tools are available to debug applications on RC systems?

The design development cycle for RC systems includes aspects of software compilation as well as hardware synthesis, as illustrated in Figure 4.2. When starting from a sequential program, this may require

1. profiling to identify compute intensive kernels
2. a rough estimate of the benefit obtained from hardware acceleration on computationally intensive kernels
3. quantification of communication costs between hardware and software to exchange data and results

In practice these partitioning steps are performed manually. Profiling of the software can be performed with well-established software tools to identify potential kernels to map to hardware. It is difficult to determine the potential benefit of hardware acceleration without actually creating the hardware

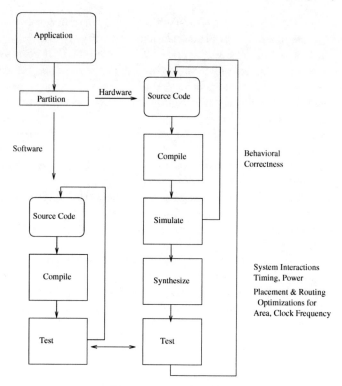

Fig. 4.2. Design Cycle

design. There have been several research efforts to provide performance estimates of hardware performance from algorithmic description. [137] developed a methodology to evaluate regular dataflow designs such as multimedia algorithms. In this work, a 23-component vector is created to characterize a design. The vector includes the I/O's, the number of arithmetic and logical operations, the degree of parallelism, and the number of iterations. The methodology was applied to six benchmarks to predict area, frequency, throughput, latency, and I/O. Area was predicted within 30% for all benchmarks, and I/O was correctly predicted on all benchmarks. However, clock frequency, throughput, and latency showed greater error.

[52] performs estimation based on execution traces of Matlab code. Again the target application area is dataflow designs in signal and image processing. The user must specify bit widths for variables. The tool builds a dataflow graph from the execution trace. The graph is annotated with estimates of FPGA resources consumed for each operation. Using a greedy algorithm, the tool explores alternative schedules of the dataflow graph with differing area/latency characteristics that meet pre-specified clock speed and execution rate constraints. Low level placement and routing are not considered in

this tool. These tools represent first efforts at algorithm-level estimation to quantify the benefit of hardware acceleration of software kernels.

The third step, understanding the I/O communication cost, is also difficult to quantify a priori. The communication cost depends on the amount of data that must be transferred between hardware and software and on the bandwidth of the communication channel. However, quantifying that cost is difficult: the application designer may be able to overlap computation with communication to hide the cost; the advertised bandwidth may differ from what is actually measured in practice; the bandwidth may be asymmetric; it may behave non-linearly in the face of congestion. All these circumstances may cause the I/O bandwidth to vary by factors of 2–4.

Once the designer has decided how to partition the application, a revised body of source code is produced which separates the source code into into hardware functionality, software functionality, and communication interfaces among them.

The hardware part of the algorithm can be expressed at a level of abstraction ranging from algorithmic to structural. It must be compiled into the target FPGA, which is a very complex, computationally intensive process. The compiled hardware code is then tested through simulation. There is usually iteration at this stage (compile–test–revise) to verify the functional correctness of the code. To generate the final hardware there are additional steps: the compiled code is then synthesized, placed, and routed to the target FPGA and tested on the hardware itself. The low level Computer Aided Design (CAD) tool stage itself has a number of sub-components and a complete sequence of steps, described in Section 4.4. Synthesis, place, and route expose issues in low level hardware timing, routing, area consumed, clock frequency achieved, power draw, and system interactions.

4.2 Languages

In this section, we discuss the problem of expressing computation that is partitioned among hardware and software components. Languages for programmable processors are algorithmic in nature, building on the Turing machine formalism of fetch, decode, and execute of a sequential stream of instructions that read and write memory. In contrast, mapping computation to reconfigurable hardware design entails fabrication of arbitrary logic circuits, exposing the maximal amount of parallelism consistent with hardware resource constraints.

For traditional hardware design, especially for circuits that must interface to external I/O devices and meet stringent timing constraints, it is desirable to use tools that mirror an abstraction of the underlying hardware resources. Modules such as shift registers, comparators, multiplexers, adders and other function units are convenient building blocks that the designer can combine, either graphically or through textual commands, to create a hierarchical, spatially parallel hardware circuit. Tools to simulate module interactions at the clock cycle level are desirable, as are tools to control and analyze the mapping, placement, and routing of the modules onto the underlying FPGA fabric.

In contrast, a reconfigurable computing problem to be mapped onto hardware is initially expressed as a sequential algorithm. There are many different ways that the algorithm can be partitioned between hardware and software, and further, many functionally equivalent hardware circuits that can be generated from the algorithm. Thus the search space in which to optimize partitioning choices, area, frequency, and throughput is very large.

Since the vast majority of FPGA applications fall in the domain of traditional hardware design, the greatest choice and capability in languages and compilers is skewed toward hardware description languages, schematic layout editors, hardware circuit simulation and synthesis. Most of these Computer Aided Design (CAD) tools are expensive ($10K's – $100K's), requiring use of high performance workstations. At the lowest level, the designer may direct the functionality and interconnect of logic blocks on the FPGA, creating dense chip-specific designs that optimize features of the particular chip being targeted. "Intellectual Property (IP)" blocks are designed at this level and often provided by the chip vendor as optimized building blocks for higher level designs. The next level of abstraction is to combine IP blocks with application-specific logic. This is called structural design. At the next level of abstraction, Register-Transfer Level (RTL), registers, function modules, control structure, and timing are all specified by the designer. Finally algorithmic and behavioral languages provide for high level functional description of computation. We now discuss each level of expression, starting with the highest level, algorithmic languages.

4.2.1 Algorithmic RC Languages

Perhaps the most straightforward approach to compiling algorithms to reconfigurable systems is to compile to the co-processor model. In this model, compute-intensive kernels within a sequential program are mapped to hardware, while the remaining program executes in software.

Using the co-processor model, hardware compilers have been developed for sequential C. The PRISM compiler by Athanas was the first to demonstrate hardware compilation of sequential C kernels for reconfigurable computing [29]. In [173], the NAPA C compiler used a pragma-driven approach to identifying compute intensive kernels, and automatically compiled the kernels to configurable logic. In [55], the design environment partitioned a sequential program between software and compute-intensive hardware kernels according to a parameterized architectural model, with automatic compilation to hardware. The compiler of [62] provided perhaps the ultimate in hardware/software co-compilation. The Garp-C compiler generated both hardware and software versions of compute-intensive kernels that were identified through automatic profiling. It also generated code to partially and dynamically reconfigure the Garp logic array so that hardware kernels could be loaded on demand during execution.

Given the opportunities for parallel processing at many levels of granularity, parallel C variants have also been investigated. Typically, acceleration of compute-intensive kernels within sequential programs yields overall speedup of 2–3X. By creating a parallel algorithm and mapping parallel processes to a combination of software and hardware, speedups of 10–100 times can be realized. This is because a parallel algorithm exposes large granularity parallelism. Within a sequential algorithm it is difficult to parallelize at more than the instruction level. In addition, because of Amdahl's Law, the overall speedup is limited by the sequential portions of the code, regardless of the kernel code speedup.

The earliest parallel C compiler was derived from the Occam communicating parallel process model with explicit parallel and sequential blocks [318], and has resulted in a commercial product Handel-C. The dbC compiler of [169] configured the FPGA as a Single Instruction Multiple Data (SIMD) parallel processor controlled by a microprocessor. The Streams-C compiler [171] supports parallel processes that can either run on hardware and software along with buffered FIFO stream communication between processes. Streams-C also includes in the language, through pragma directives, the notion of multiple named memory banks, and allows the programmer to map arrays to memory banks. Streams-C has been commercialized into the Impulse C product. The Streams-C compiler is also available in open source code on the Los Alamos reconfigurable computing web site [248]. The SA-C of [54] was a single assignment variant of C optimized for image processing operations. The SA-C language introduced the notion of a convolution window, and the compiler generated a bank of shift registers to provide efficient pixel neighborhood access for local convolutions over a sliding window. The MAP compiler from SRC Computer (see Chapter 3 Section 3.2.3) translates C and Fortran to SRC-specific hardware.

Java compilers have also been developed, such as the Forge compiler from Xilinx, and the SC compiler [397] from BYU. The Forge compiler generates hardware from sequential Java bytecode. The SC compiler, like Streams-C, provides a parallel processing model with channel communication between processes. The Java bytecode of each process in a program is synthesized to hardware, and FIFO communications channels may be read or written by processes.

In the signal processing domain, a Matlab compiler was introduced by [192] which has been also been commercialized. Another project from Leiden University translates static nested loop programs written in Matlab into Kahn Process Networks. Communications channels between processes are synthesized and the process bodies invoke pre-built IP cores ([383], [440]). At the Simulink level, Xilinx Corp. markets the System Generator tool to map Simulink blocks onto modules from the System Generator library. The Viva graphical programming environment [382] is another schematic-oriented visual interface for composing hardware blocks.

Algorithmic Language Example

Figure 4.3 shows a small algorithmic description in Streams-C [159]. This module is the first step of a phase modulation sorter that locates binary phase shift keying (BPSK) signals in wide-band data. The example illustrates many attributes of algorithmic RC languages. The initial "pragma SC memory" lines assign arrays to memories. Reconfigurable computing systems offer a rich collection of memory types, such as SRAM or DRAM external to the chip, on-chip RAM blocks, or even logic blocks configured as memories. This example uses external memory "mem_1" and configurable logic block memory "DP_FFT_1"

The first command in the outer loop uses a synchronization construct, the "sc_wait" intrinsic, which allows a process to signal a condition to another process. In this case, the bpsk process waits for the completion of an FFT module that performs four FFTs and returns the result into "sum." The next line calculating the "Threshold" uses the C type casting mechanism to specify a 24-bit integer. Since there isn't a defined word size in configurable logic, RC algorithms often use non-standard integer sizes, and RC languages extend the underlying type system with additional integer sizes. Calculation of the threshold uses a divide by a power of two and a multiply by a power of 2. Using a power of two for multiplies and divides means that the compiler can convert an operation that is expensive in configurable logic into a simple shift operation.

The next pragma asks the compiler to pipeline the loop over "L." Loop pipelining allows multiple loop iterations to be overlapped, increasing parallelism and decreasing loop execution time. The next loop checks the input data (read into "squared_data" from the input stream) for the value 1 cast as a 2-bit integer. If it is a 1, the array "data_out" is indexed in six different frequency bins, and all six bins are concatenated together to form a 12-bit integer "check'." If this is non-zero, the event is recorded as the "k'th" event in an array and the counter "k" is incremented.

In Section 4.3 we will see how the compiler synthesizes this algorithm into a Register-Transfer-Level hardware description.

4.2.2 Hardware Description Languages (HDL)

Hardware description languages can be used to express computation at the behavioral, RTL, structural, or even device specific level.

Behavioral hardware description languages allow the designer to describe the behavior of a circuit, from which many alternative circuit realizations can be generated. While behavioral languages share many similarities with

```
#pragma SC memory mem_1 data_out
#pragma SC memory mem_1 event_data
#pragma SC memory DP_FFT_1 x

for(j=0; j<Q; j++){
// external IP core generates 4 ffts
// when fft IP core 'finishes' processing,
// it returns the sum of 4 ffts
  sum = sc_wait(input_signal);

// calculate the threshold
Threshold = (sum/(sc_int24)2048)*(sc_int24)4;

  for(i=0; i<L; i++){
#pragma SC pipeline
  if (x[i] < Threshold )
   data_out[i] = 0;
  else
   data_out[i] = 1; //strong value
 }
// check for BPSK
// if squared_data = 1 check six frequency bins of the
// non squared data for a 1
  for(j=0; j<L; j++){
   squared_data = sc_stream_read(input_stream);
   if (squared_data == (sc_int2)1){
    tmp0 = data_out[j];
    tmp1 = data_out[j/(sc_int32)2 + (sc_int32)1];
    tmp2 = data_out[j/(sc_int32)2 - (sc_int32)1];
    tmp3 = data_out[(FFTSize-j)/(sc_int32)2];
    tmp4 = data_out[(FFTSize-j)/(sc_int32)2 + (sc_int32)1];
    tmp5 = data_out[(FFTSize-j)/(sc_int32)2 - (sc_int32)1];
    check = sc_catenate(tmp5,tmp4,tmp3,tmp2,tmp1,tmp0);
    if ((sc_int12)check != (sc_int12)0){
      event_data[k] = (sc_int32)j; //peak detection
      k++;
    }
```

Fig. 4.3. Phase Modulation Sorter Excerpt in Streams-C

algorithmic languages described in Section 4.2.1, the term "behavioral language" generally refers to the high level modeling and simulation constructs of existing hardware description languages such as VHDL and Verilog. Like high level language RC compilers, behavioral compilers synthesize datapath and control state machines from algorithmic source code, resolving clock cycle level timing and application-specific micro-architecture. Behavioral languages include many features of algorithmic languages including function call, looping constructs, and a sequential instruction stream. The synthesizable subset of behavioral HDLs usually does not include hardware synthesis of generalized pointer references (e.g., function pointers) or dynamic memory allocation.

Using the RTL or structural level of hardware description language, the designer can control the selection, instantiation, and interconnection of hardware modules. The designer can specify the instruction level parallelism within the circuits, direct the unrolling of loops, and manually pipeline a design. Hardware description languages include VHDL, Verilog, System C, and Handel-C.

VHDL and Verilog are mature, industry standard HDLs, with many vendors offering simulation and synthesis tools. Behavioral, RTL, and structural levels of description can be used inter-changeably in these languages. System C is a C++-based library used for modeling system level behavior. As the base language is C++, software processes can be more easily modeled than in a more traditional HDL. Synthesis tools for System C are emerging, but do not approach the maturity of VHDL or Verilog synthesis products. Handel-C is also a relatively new product in comparison to VHDL or Verilog. Derived from the hardware Occam compilation research effort from Oxford University [318], Handel-C follows the Communicating Sequential Process (CSP) model. Handel-C requires the designer to explicitly delineate parallel processing blocks within a process. It includes intrinsics for inter-process communication, as does System C 2.0.

A VHDL Example

To illustrate characteristics of hardware description languages, we show in Figures 4.4 and 4.5 a portion of RTL VHDL for the BPSK example. The code was generated by the Streams-C compiler. The code fragment in Figure 4.4 starts with an "entity" declaration, preceded by references to libraries used by the entity. Certain standard IEEE libraries for standard logic and numeric standard data types and operations are provided with VHDL compilers. Designers may also include their own libraries. In this example, two additional libraries "pipeControl" and "strmshift" in the current library ("work") are also included. The purpose of the entity declaration is to define the I/O interface of the module. Each port may be either input, output, or both. The data type of each port must be defined. In this example, there are several control lines entering the module, such as Clk, Reset and various handshake signals, as well as interfaces to the FFT, a memory, and an input stream of data. The 9-bit "Instruction" input from a controller tells the datapath which instruction to execute.

In VHDL, there may be many different **architectures** that implement an entity. Figure 4.5 shows a fragment of the RTL architecture for BPSK. Within the architecture, registers may be defined using the **signal** keyword. A VHDL **process** appears next. The process's parameters are called its sensitivity list. Each time one of the parameters changes, the process is activated. In this example, the process is activated at each change to the clock signal or the reset signal. Within the process body, the bulk of the work is done on the rising edge of the clock. The code fragment shows a portion of the BPSK pipeline, which is performed in Instruction 6. There are many pipe stages; the code shows only 4 of them. Within a pipe stage, all signal assignments occur concurrently during the clock period. At the end of the clock period the signal assignments are complete.

Figure 4.6 shows how structural components are expressed in VHDL. The top level BPSK entity has a "Structure" architecture. There are two components, which must be declared first, and whose port list must match their entity declarations. For example the component bpsk_dp's port list as declared within the Structure architecture must be the same as was declared in the bpsk_dp entity declaration in Figure 4.4. In this example, there is a second component, a sequencer bpsk_seq. The sequencer component declaration illustrates the use of generics in VHDL. The instruction width InstWidth is a parameter to the sequencer entity. When the sequencer is instantiated, the "generic map" assigns constant values to the generic parameters. In the example, the width of the Instruction bus is 9. When each component is instantiated, there is a "port map" that assigns names defined within the architecture Structure to ports of the component. In this case, the same names have been used for the clock, reset signal, etc. in Structure as were used for the port names in the bpsk_dp and bpsk_seq entities.

VHDL is a large and complex language. The intent of this example is to illustrate a few VHDL concepts (entities, architectures, ports, processes, components). The reader is referred to VHDL texts such as [28] for a comprehensive language tutorial.

The lowest level of description actually specifies and instantiates logic blocks on the FPGA. This level is typically used by the FPGA vendor who provides highly optimized hardware macros for use as building blocks. Such a building block, a logic block configured as a random access memory, is invoked in Figure 4.7, in the Electronic Design Interchange Format (EDIF) .

4.3 High Level Compilation

Compiling for reconfigurable computers is considerably more difficult than compiling for conventional processors. With conventional processors, the instruction set architectures (ISA) is given. The problem is to map an abstrac-

```
LIBRARY IEEE;
USE IEEE.std_logic_1164.ALL;
USE IEEE.numeric_std.all;
use work.pipeControl.all;
use work.strmshift.all;

ENTITY bpsk_dp is

    port (
        Clk      : in std_ulogic;          -- System clock
        Reset    : in std_ulogic;          -- System reset
        NewInst  : in std_ulogic;          -- New instruction issued
        iBool    : out unsigned(0 downto 0); -- instruc cond flag
        iDone    : out unsigned(0 downto 0); -- instruc done flag
        DP_FFT_0_MAR: OUT unsigned(11-1 downto 0);
        DP_FFT_0_MDR_I: IN std_logic_vector(24-1 downto 0);
        DP_FFT_0_MDR_0: OUT std_logic_vector(24-1 downto 0);
        DP_FFT_0_W_EN: OUT unsigned(0 downto 0);
        DP_FFT_0_R_EN: OUT unsigned(0 downto 0);
        DP_FFT_0_Stall: IN unsigned(0 downto 0);
        mem_1_MAR: OUT unsigned(32-1 downto 0);
        mem_1_MDR_I: IN std_logic_vector(64-1 downto 0);
        mem_1_MDR_0: OUT std_logic_vector(64-1 downto 0);
        mem_1_W_EN: OUT unsigned(0 downto 0);
        mem_1_R_EN: OUT unsigned(0 downto 0);
        mem_1_Stall: IN unsigned(0 downto 0);
        input_stream_data : IN std_logic_vector(1 downto 0);
        input_stream_en: OUT unsigned(0 downto 0);
        input_stream_err: IN unsigned(0 downto 0);
        input_stream_open: OUT unsigned(0 downto 0);
        input_stream_close: OUT unsigned(0 downto 0);
        input_stream_rdy: IN unsigned(0 downto 0);
        input_stream_eos: IN unsigned(0 downto 0);
        input_signal_data: IN std_logic_vector(23 downto 0);
        input_signal_rdy: IN unsigned(0 downto 0);
        input_signal_ack: OUT unsigned(0 downto 0);
        output_signal_data: OUT std_logic_vector(31 downto 0);
        output_signal_en: OUT unsigned(0 downto 0);
        standard_initiate_insignal_data:
                IN std_logic_vector(0 downto 0);
        standard_initiate_insignal_rdy: IN unsigned(0 downto 0);
        standard_initiate_insignal_ack: OUT unsigned(0 downto 0);

        Instruction : in unsigned(8 downto 0);
        Stall   : out std_logic
        );
end bpsk_dp;
```

Fig. 4.4. BPSK Entity Declaration in RTL VHDL

```vhdl
architecture RTL of bpsk_dp is
    signal sum_p5: signed(23 downto 0);
    signal Threshold_p6: signed(23 downto 0);
    signal j_p11: signed(31 downto 0);
    signal k_p12: signed(31 downto 0);
    signal squared_data_p13: signed(1 downto 0);
    signal tmp0_p14: unsigned(1 downto 0);
    ...
  begin

  dp_clocked_process: process(Clk, Reset)
  begin
  if   (Reset = '1') then
    ...
  elsif rising_edge(Clk) then
    ...
   if Instruction(16#6#) = '1' then
     iDone <= PipeDone_i;
     if pipeEnable_i(16#0#) = '1' then
       squared_data_p13 <= signed(input_stream_data);
     end if;
     if pipeEnable_i(16#1#) = '1' then
       if unsigned(squared_data_p13) = unsigned(to_signed(1, 2)) then
         suif_tmp8_p48 <= to_unsigned(1, 1);
         else suif_tmp8_p48 <= to_unsigned(0, 1);
       end if;
     end if;
     if pipeEnable_i(16#2#) = '1' then
      if suif_tmp8_p48 =  1 then
       suif_tmp_p39 <=  sc_asr (j_p11,
               to_integer(to_unsigned(1, 32)) );
      end if;
      if suif_tmp8_p48 =  1 then
       suif_tmp14_p54 <= to_signed(4096, 32) - j_p11;
      end if;
     end if;
     if pipeEnable_i(16#3#) = '1' then
      if suif_tmp8_p48 =  1 then
       suif_tmp3_p43 <= suif_tmp_p39 + to_signed(1, 32);
      end if;
      if suif_tmp8_p48 =  1 then
       suif_tmp4_p44 <= suif_tmp_p39 - to_signed(1, 32);
      end if;
      if suif_tmp8_p48 =  1 then
       suif_tmp0_p40 <=  sc_asr (suif_tmp14_p54,
                  to_integer(to_unsigned(1, 32)) );
      end if;
      if suif_tmp8_p48 =  1 then
       suif_tmp11_p51 <= signed(mem_1_MDR_I(31 downto 0));
      ...
      end if;
     end if;
   end if;
  end if;
  end process;
```

Fig. 4.5. BPSK Pipeline in RTL VHDL

```
ENTITY bpsk_top is
    port (
        Clk      : in std_ulogic;        -- System clock
        Reset    : in std_ulogic;
        ...
    );

architecture Structure of bpsk_top is

COMPONENT bpsk_dp
      port (
      ...
      );

COMPONENT bpsk_seq
      generic (
          InstWidth : Positive        -- Instruction width in bits
          );
      port (
...
          );
          ...
  begin  -- Structure

  seq: bpsk_seq
  generic map (
          InstWidth => 9)
        port map (
          Clk => Clk,
          Reset => reset,
          NewInst => NewInst,
          iBool => iBool,
          iDone => iDone,
          Inst => inst);

    dp: bpsk_dp
        port map (
          Clk => Clk,
          Reset => reset,
          ...
        );
end Structure;
```

Fig. 4.6. BPSK Structure Instantiation in VHDL

```
(cell RAM16X1D (cellType GENERIC)
    (view view_1 (viewType NETLIST)
        (interface
            (port D (direction INPUT))
            (port WE (direction INPUT))
            (port WCLK (direction INPUT))
            (port A0 (direction INPUT))
            (port A1 (direction INPUT))
            (port A2 (direction INPUT))
            (port A3 (direction INPUT))
            (port DPRA0 (direction INPUT))
            (port DPRA1 (direction INPUT))
            (port DPRA2 (direction INPUT))
            (port DPRA3 (direction INPUT))
            (port SPO (direction OUTPUT))
            (port DPO (direction OUTPUT))
        )
    )
)
```

Fig. 4.7. Instantiating a RAM Block, EDIF format

tion of the ISA as represented by a high level programming language onto a concrete ISA. With superscalar architectures particularly, much of the optimization occurs at run time in the micro-architecture that implements an ISA, simplifying the job of the compiler.

In contrast, the FPGA has no instruction set architecture. The task of the algorithmic/behavioral compiler is to devise a micro-architecture customized to the specific application, including datapath (the arithmetic units and registers), memory hierarchy, I/O, and sequencer (to control the sequence of datapath operations, memory access, and I/O). From the data types used and the operations within a program, the compiler must generate function units to execute the primitive operations. Often there is a very large module library from which to select function units: should the adder be 8-, 16-, 32- or 64-bit? Fixed precision, fixed point, floating point? Should a serial, pipelined, or parallel implementation be chosen? Trade-offs between area and clock speed affect pipeline generation strategy.

Given the arrays and other variables in a HLL program, the compiler must decide where each variable resides. As seen in Chapter 3, reconfigurable computing systems include a complex memory hierarchy ranging several orders of magnitude in size and latency, making the trade-off space very large. Once variables have been assigned locations, the compiler must generate hardware to read and write memory to/from on-chip registers and function units. High speed I/O such as a data stream from an A/D converter, imposes hard real-time constraints on the design.

The combination of these constraints and choices makes compiling algorithmic languages to reconfigurable systems a very difficult multi-objective combinatorial optimization problem. Equally difficult tasks are required for RTL synthesis to logic gates, mapping gates onto configurable logic blocks, placing virtual logic blocks to physical resource, and routing among the logic blocks.

In this section we discuss high level compilation. RTL synthesis and low level mapping, placement and routing are described in Section 4.4.

4.3.1 Compiler Phases

Fig. 4.8. Steps of Algorithmic/Behavioral Synthesis

Figure 4.8 outlines the compilation steps from algorithmic code to RTL level. The first phase simply parses the high level language syntax into an Intermediate Representation (IR) that includes symbol tables to store the program's variables and data types, and an Abstract Syntax Tree (AST) representation of the executable code. In the analysis/optimization phase, control and data flow graphs are created for each subroutine. Data and control dependencies are identified among statements within a subroutine. Optimizations are then applied to the graph, which may require additional analysis. The scheduling phase uses the optimized flow graphs to define the datapath and associated control, typically at a subroutine granularity. The scheduled graph is then output to a target language, often a RTL/structural HDL.

We next discuss in more detail the analysis/optimization and scheduling phases of compilation.

4.3.2 Analysis and Optimizations

The analysis phase creates a control/data flow graph representation of each subroutine. The control flow graph captures the sequence of operations. For example, if-statements and loops are converted into a graph whose edges model the flow of control through the conditionals and branches. The nodes of the control flow graph are data flow graphs that model the data dependencies of basic blocks, the straight line code segments. Many compilers use a Static Single Assignment (SSA [23]) representation, in which re-assignments to the same variable within a block are replaced by assignments to new temporaries. At the end of the block, a "phi-node" is inserted for each variable that was re-assigned in the block. The phi-node selects among all possible re-assignments to the variable, so that the correct value is propagated out of the block. The use of SSA exposes additional instruction level parallelism into the data flow graph.

Another technique to expose additional instruction-level parallelism is called if-conversion. Rather than representing an if-statement with a control flow graph, the if-statement is converted to a straight-line sequence of guarded (or "predicated") statements. The guard is the if-condition for all statements within the true-branch of the if-statement, and the negation of the if-condition for the else-branch statements. When there are nested if-statements, the if-condition at the inner levels of nesting can become quite complex. Often a Binary Decision Diagram (BDD [402]) format is used to express complex conditional expressions.

Once the AST has been translated into a control/data flow graph in a format that exposes instruction level parallelism, standard compiler optimizations can be performed ([301]). Optimizations are semantics-preserving transformations that (hopefully) improve the quality of the generated hardware. These optimizations,examples of which are shown in Figure 4.9, include constant propagation, common sub-expression elimination, strength reduction, dead code elimination, and code motion to move loop invariant code outside the loop.

Another parallelism-exposing transformation is loop unrolling, as shown in Figure 4.10. Loop unrolling, in conjunction with other transformations such as constant propagation, dead code elimination, and the transformation of the arrays A and B into register banks, can transform the loop into four simple assignment statements that can occur in parallel.

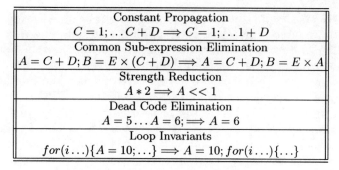

Constant Propagation
$C = 1; \ldots C + D \Longrightarrow C = 1; \ldots 1 + D$
Common Sub-expression Elimination
$A = C + D; B = E \times (C + D) \Longrightarrow A = C + D; B = E \times A$
Strength Reduction
$A * 2 \Longrightarrow A << 1$
Dead Code Elimination
$A = 5 \ldots A = 6; \Longrightarrow A = 6$
Loop Invariants
$for(i \ldots)\{A = 10; \ldots\} \Longrightarrow A = 10; for(i \ldots)\{\ldots\}$

Fig. 4.9. Standard Compiler Optimizations

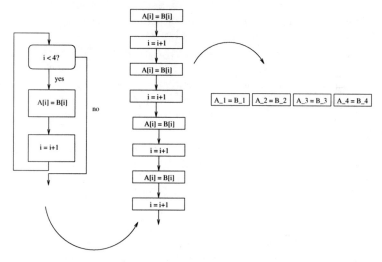

Fig. 4.10. Loop Unrolling

4.3.3 Scheduling

Scheduling is the process of assigning an ordering, clock cycle by clock cycle, to the operations of the control/data flow graph. The earliest that an operation can be scheduled is after all operations on which it depends (either in control flow or data dependence) have been scheduled (As Soon As Possible – ASAP – scheduling). Ideally it should be scheduled before operations that depend on it could initiate (As Late As Possible – ALAP – scheduling). . The classical technique of force-directed scheduling [322] attempts to find the best time to schedule an operation within that window.

The scheduling of an operation also depends on the number of clock cycles required by that operation to complete. Many operations, such as comparisons or small adders, complete within a clock cycle. Other operations take multiple clock cycles. Often multi-cycle operations are pipelined, so that new operands can be introduced every clock cycle, and after an initial latency, a result is available every cycle. The instruction selection phase within the scheduler chooses among available alternative implementations of the operators.

To complicate the scheduling process even further, the length of a clock cycle is not a single fixed time for each different sort of FPGA. The flexibility inherent in the FPGA architecture makes it possible to create circuits that range from 5 up to 500 MHz. For a given design, the amount of work scheduled within a clock cycle determines the clock cycle length. Unfortunately, the final clock cycle length is only determined in the last phase of physical synthesis during Place and Route. At the relatively high level of the compiler in which scheduling of behavioral code occurs, heuristics must be used to select operators such that appropriate trade-offs are made between clock frequency, area, and throughput. Often directives from the programmer or circuit designer are used to help navigate this complex optimization space.

The most important optimizations a reconfigurable computing compiler can perform for FPGA-based designs are instruction level parallelization, using transformations discussed above, and pipelining. This is because attainable clock frequency on an FPGA lag modern microprocessors by at least a factor of ten. Spatial parallelism must be exploited to compensate for the slow clock speed.

When instruction level parallelism is exposed, the compiler can schedule many operations in parallel. Similarly, when a sequence of operations is pipelined, many levels of the pipeline operate in parallel. Pipelining can occur within individual operations, as for example, with a pipelined multiplier or floating point unit. Pipeline stages can be inserted by the compiler in order to increase clock rate – when fewer interdependent operations are scheduled in the same clock cycle, the frequency can be increased. The trade-off is in area: pipelining introduces registers to hold intermediate values.

Pipelining can also be applied at the loop level. The goal of loop level pipelining is to initiate a new loop iteration before the previous iteration is complete. It is of course desirable to start a new iteration every clock cycle. This may be possible if the loop iterations are independent and enough delay registers are introduced. Iterative modulo scheduling [249] is a well known technique for pipelining loop iterations.

4.4 Low Level Design Flow

The purpose of low level design flow is to pass the register transfer level or structural code through a Computer Aided Design (CAD) tool chain that ultimately generates the FPGA's configuration bit stream (see Figure 4.11).

4.4.1 Logic Synthesis

The first step in that sequence is logic synthesis, which translates the register-transfer level description of a hardware design into an optimized gate level representation. Since the mid-eighties, logic synthesis has emerged as an essential part of the CAD tool suite, and is the subject of many texts, e.g., [289]. A brief overview of the logic synthesis function follows.

In logic synthesis, two broad categories of digital circuits may be synthesized, combinational or sequential. The outputs of a combinational circuit depend only on current inputs. Multiplexers, decoders, and boolean equations are examples of combinational logic. A sequential circuit requires feedback, as it retains state after the inputs have been removed. The state is updated based on a clock. Latches and flip-flops are the building blocks of sequential circuits, from which registers, counters, and state machines can be constructed.

For combinational circuits, the logic synthesis problem is to

- generate a set of boolean equations from the RTL/structural specification
- transform the equations into two level (sum of products or product or sums) logic or multi-level logic
- minimize the two (multi) level circuit relate to a cost model.

Sequential circuit synthesis is concerned with finite state machines. The logic synthesis algorithms must

- specify or identify state machines in the RTL/structural description
- minimize the number of states
- encode the states in a compact binary representation
- optimize the resulting two (multi) level logic

The challenge of logic synthesis is to optimize trade-offs in area, speed, and power in order meet design constraints of the system. Optimization can occur at the technology-independent level or be targeted to a particular implementation technology.

Technology independent optimizations define an area cost model based on the number of literals in the set of boolean equations, and a delay cost model based on the length of the longest dependency chain in the set of equations. They attempt to reduce redundant logic and common sub-expressions. Technology dependent optimization use the cost models associated with a particular library of mappings from general logic gates into a set of pre-defined modules.

Technology dependent optimizations are often performed during the technology mapping phase in which the general logic gates are mapped to k-input Look Up Tables (LUT), the basic logic cell of FPGAs. Conventional technology mapping uses a limited set of library cell elements. Each cell has a cost, and optimization algorithms look for optimal covering of the general logic gates onto the library cells. However, since a k-input configurable logic block can be configured into an arbitrary boolean function of k inputs, the conventional approach would lead to a very large library of cells. Therefore, algorithms specifically designed to map gates onto k-input LUTs have been developed (see, for example, [156]).

Once a design has been expressed in terms of a library specific to an FPGA, the next several stages of the design flow include technology mapping, logic placement, the routing of connections among the physical logic resources, and programming data generation. Figure 4.11 illustrates this low-level design flow. Though a detailed description of these particular design operations are out of the scope of this book, the remainder of this section will describe each phase in general terms.

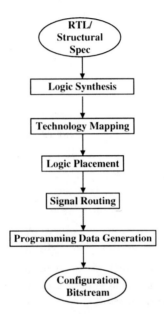

Fig. 4.11. Low-Level Design Flow for FPGA Design Mapping

For a detailed discussion of this process, we recommend [50], which provides an excellent overview of the subject and then provides considerable details on each design flow step from technology mapping to routing. Further, they describe how to represent FPGA resources for use in CAD algorithms, how to estimate circuit timing and costs, and how they implemented their own FPGA tool framework (VPACK, T-VPACK, VPR) and FPGA architectural exploration tools—tools that continue to be used by academia [430] and the commercial sector [260, 261].

4.4.2 Technology Mapping

When a design has been synthesized automatically or manually into a set of design primitives that the low-level FPGA tools understand, the first phase of converting a design expressed in these primitives into programming data for a particular reconfigurable logic device is called technology mapping. The technology mapping tools convert the netlist of library primitives into a netlist of physical device resources that efficiently implement equivalent functions.

The technology mapping approaches vary depending on the specific architecture used by a reconfigurable logic device, but a commonly used approach used for FPGAs is as follows. First, conventional logic optimization techniques are used to prune redundant logic from a design as well as to simplify the logic. Even though some level of optimization may have been performed during design synthesis, frequently additional optimization may be possible, especially, if multiple design units that have been independently synthesized are being integrated together. For instance, a pre-synthesized module that performs many functions may be hard-wired to perform a single function for a given design, often resulting in redundant logic that can be optimized away by the low-level FPGA tools.

Next, the mapping software uses an algorithm to cover the netlist of library primitives with a set of logic resources that perform an equivalent function. Several effective algorithms for performing the netlist covering operation using four-input LUT logic elements have been developed, including Chortle [158], Chortle-crf [157], DF-Map [88], FlowMap [87], and DAG-Map [232]. These algorithms consider how to pack logic functions in LUTs based on constraints such as area, circuit speed, and algorithm speed. As a recent article [233] by Kao and Lai illustrates, research in the area of technology mapping continues to progress.

Finally, the tools perform a clustering process on the netlist of FPGA physical resources to determine how to map the used resources to the larger clusters of logic found in FPGA logic blocks. This is done to maximize the utilization of the FPGA logic blocks as well as to take advantage of internal logic block routing and thus minimize the use of slower, less efficient general programmable routing structures. As with LUT mapping, a significant amount of research has been and continues to be performed in this area [10, 89–91, 191, 354]

4.4.3 Logic Placement

During the next phase, logic placement, CAD tools determine an efficient placement for each mapped logic block among all of the possible physical locations for that block within the reconfigurable logic device. This placement can be driven by constraints such as minimizing wire length, ensuring routability within the array, and/or maximizing circuit performance (i.e., timing-driven placement). As discussed in [50], min-cut, analytic, and simulated annealing approaches are commonly used to perform placement.

Simulated annealing [361] appears to be the algorithm of choice for most FPGA implementation tools [388]. With this method, the algorithm is given an initial placement and then swaps the placement of pairs of logic blocks, evaluating the swaps using some predefined cost function. With simulated annealing, an annealing schedule is used to determine the rate at which more costly moves will be accepted as a way of avoiding local minima. As the schedule progresses in time, fewer and fewer moves that lead to a higher cost will be accepted. This technique produces good placement quality at the expense of long execution times [387]. Further, the technique does not take advantage of circuit hierarchy to improve placement. In [352], a recursive clustering processes performed before annealing improved executing time, but, again, the natural design hierarchy was not used.

In an effort to improve design time through design reuse, many FPGA designs have started to use pre-defined macros, but simple simulated annealing techniques do not effectively deal with these circuit blocks. Tessier developed a placement system for the Frontier tool [388] that performs floorplanning of macro blocks through clustering and bin placement techniques and then uses low-temperature simulated annealing on individual macro blocks only if routability or circuit performance problems exist.

4.4.4 Signal Routing

The final step in low-level FPGA design implementation is routing the signals between inputs and outputs of the placed logic blocks by appropriately configuring pass transistors, buffers, and multiplexers. As pointed out in [50], two generals styles of routers exist. The first style performs the routing process in two steps. First, a global routing process is performed which determines the logic block pins to be used for a net and the routing channels to be used. A detailed routing process—the second step—then selects the actual wires used within the routing channels for each net.

The second general style performs both global and detailed routing as a single combined step. [50] suggests that this second style of routing is more appropriate for FPGAs since it tends to avoid the constraints that a separate detailed routing step encounters. Unlike ASICs, FPGAs' routing resources are predefined so the detailed routers have limited flexibility in how to perform routing within the FPGAs' routing channels chosen by the global routers.

Routers can be driven by different constraints. Some of the more successful algorithms tend to consider both timing and routing congestion constraints. Many of these routers internally have some form of maze router [253] which implements Dijkstra's shortest (or minimum cost) path algorithm [123] to route between logic resources where the costs of each route are related to some balance of timing and congestion or routability constraints. The PathFinder algorithm [133] is an excellent example of such a router and has been used as the basis other routing algorithms, including the one used in the VPR timing-driven router described in [50]. With this router, timing critical nets are allowed to take the shortest paths despite congestion. Less critical nets are forced to take less congested paths. During each iteration of the algorithm, all of the nets are ripped up and rerouted in the same order, but the costs assigned to each net are modified to balance the needed speed of the connection with the congestion found in the routing resources. The router completes when all routes are valid and timing constraints have been met.

4.4.5 Configuration Bitstreams

The final step in the low-level design flow is the generation of the actual data used for programming the FPGA or other reconfigurable logic device. These configuration bitstreams, as described in Chapter 2, dictate how the various resources are configured—turning off or on pass gates, turning off or on routing buffers, selecting specific multiplexer inputs, determining the contents of LUTs and RAMs, etc. The programming data itself may also contain checksums or CRC values for checking data integrity and, in the case of Xilinx and possibly other families, commands for controlling the configuration process and parameters are intermingled with the data.

Though the above function is not challenging when compared with the previous steps of the design flow, some interesting opportunities exist for tools that can create and manipulate configuration bitstreams. An excellent example of this are the JBits tools and application programming interface (API) [186] created by Xilinx for their XC4000, Virtex, and Virtex-II FPGA families. The JBits API allows Java programmers to generate bitstreams by directly specifying how individual resources can be configured. For instance, the values for a specific LUT could be set, a specific input to a multiplexer could be selected, buffers and pass transistors could be turned on or off, etc. For Virtex, Xilinx also provided an additional API on top of JBits called JRoute [235] that would allow Java programmers to automatically create and remove connections between logic inputs and outputs. Though not a sophisticated, timing-driven router, the JRoute router greatly simplified the creation of circuit designs when operating at the low-level of abstraction provided by JBits.

Besides providing a way to create new bitstreams or modify existing ones, JBits also provided a method for creating partial bitstreams for Virtex and Virtex-II, enabling designers to craft hardware designs that could be dynamically reprogrammed during execution. For instance, with JBits, constants located in LUTs or Block RAMs could easily be changed to other values by loading the chip with a series of partial bitstreams that affected only those areas of the chip—a method employed in [321] for changing keys or modes in a high-speed DES encryption/decryption module implemented in a Xilinx Virtex FPGA. Likewise, parts of circuits can be added or removed while the hardware is executing. One of the more exotic uses of run-time reconfiguration and JBits that actually adds and removes circuitry is described in [269], where the temperature of the FPGA die was being monitored by dynamically configuring, operating, and then removing individual instances or arrays of specialized thermal sensor circuits after executing some function on the

FPGA. JBits has also been coupled with evolutionary algorithms to evolve circuits in hardware (e.g., [250, 259, 268, 367]).

4.5 Debugging Reconfigurable Computing Applications

Like any other hardware or software system, reconfigurable computing applications will often fail to operate as expected. Traditional symbolic debuggers and other software tools can be used to debug problems with the software portion of a reconfigurable computing application, but the hardware portion of the application will require different tools. Over the short history of reconfigurable computing, several significant, though largely unnoticed, efforts have striven to provide some level of debugging support for reconfigurable computing applications. We will provide only a brief overview of these efforts. The reader is encouraged to refer to [182] for a more complete overview.

4.5.1 Basic Needs for Debugging

Several features are required to provide a productive debugging environment for hardware. These include:

Observability: The ability to determine the values of signals and state elements in a circuit.

Controllability: The ability to set the state of circuit to a desired value.

Execution Control: The ability to control the design's clock or clocks as well as control the design's inputs.

Observability provides the designer with the current state of the circuit so its operation can be understood. Controllability allows the designer to force the circuit into specific states so the designer can experiment with how the circuit behaves as well as allowing checkpointing for restarting the circuit at specific points in its execution. Execution control is important for efficient debugging since it can provide the designer with precise, interactive control over how many cycles a design executes, making it possible for the designer to more readily localize design errors temporally. Another possible form of execution control, hardware breakpointing, can allow the designer to stop a design's execution when certain conditions occur that may be related to a design problem.

In addition to these specific features, the effectiveness of a debugging system for hardware also depends on several other features including:

Debugging Data Bandwidth: This affects how much design state can be transferred between the reconfigurable computing system and its host. As an example of how this can be important, consider a system with with poor debugging data bandwidth. With this system, very little design state can reasonably be observed without significantly slowing system execution.

System Execution Speed: This quality is a relevant issue since it clearly impacts how long the debugging task will take to complete as well as how realistic the execution environment appears to the application being debugged.

Instrumentation Costs: These refer to the costs in terms of extra hardware required for providing the observability, controllability, and execution control features to the debugging system. This instrumentation can be done within the reconfigurable logic device's design, within the application's design itself, or at the configurable computing system level. When the devices and systems do not provide enough observability, controllability, and/or execution control, the configurable computing application itself can be instrumented to provide improvements in these areas, but often at significant cost in terms of circuit area and/or speed.

Ease of Use: Ease of use for debugging features must be high, otherwise, a designer will not be very productive. This includes automating the mapping of hardware state to the designer's hardware design and displaying the information in a intuitive way. Automation can also help the designer by automatically instrumenting circuits to improve some debugging characteristic, significantly reducing the designer's burden.

4.5.2 Debugging Facilities

A number of devices, systems, and design tools used in reconfigurable computing have included some support for debugging. This section will provide a brief overview of this work, focusing mainly on features of devices, systems, and design tools that provide improved observability, controllability, and/or execution control to the designer. Again, refer to [182] for a more complete discussion of many of these debugging helps.

Devices

With regards to devices and observability, two main features are worth noting. First, the boundary scan feature (IEEE Std. 1149.1) provided in many of todays ICs can provide observability for the I/O pins of devices. Often referred to as JTAG for the people who defined the standard (the Joint Test Action Group), FPGA vendors have extended the interface to allow for configuring FPGAs devices, reading back this programming data once the device is programmed, and for providing another interface for interacting with the device's user logic. The JTAG interface to user logic has become a common way for logic analyzers embedded within FPGA circuits to communicate with external equipment [12, 423] (see the "Hardware Debugging Tools" section below). JTAG and special debugging interfaces have been added to FPGAs to support debugging of the on-chip embedded processors as well [426, 428].

One of the most beneficial observability features built into FPGA devices has been Xilinx's configuration readback feature. Not only does it allow the device's programming data to be read out but it also can be used to sample the state of RAMs and flip-flops within the devices.

With regards to controllability, the JTAG interfaces allow the I/O signals to be forced to certain states, but this is fairly limited since most of the flip-flops and signals reside internal to the design. The device programming or configuration interface provides some level of controllability for on-chip RAMs for most FPGAs. Flip-flop state can be altered using configuration and flip-flop resets in most FPGAs, but the ability of Virtex-II [424] and later Xilinx FPGAs to initialize flip-flop state independently from the set-reset logic provides a significantly improved degree of controllability for FPGAs since the design itself does not have to be altered in some way to force flip-flops to a chosen state.

Though devices do not directly provide execution control facilities, some FPGAs provide clocking resources that can aid in debugging. For instance the PLLs, DLLs, and DCMs of Altera's and Xilinx's latest FPGAs can be employed in on-chip circuits to provide clock single-stepping or even hardware breakpointing [182]. Also, the introduction of clock buffers in Virtex-II that allow clocks to be stopped in a clean, predictable manner is helpful. Additionally, on-chip clock multiplexers on Virtex-II and later FPGAs, which allow for clean switching between clock sources on chip, can potentially be used to test a design with multiple clock sources or change back and forth from a continuously running clock to a clock that can be operated under single-stepped control.

Reconfigurable Computing Systems

Several reconfigurable computing systems have provided some level of debugging support for observability, controllability, and/or execution control. The level of support in most reconfigurable computing (RC) systems, though, has been minimal, adding to the challenges of developing and deploying reconfigurable computing applications. The next several paragraphs describe the debugging support provided with several of these systems.

RC Systems with Some Debugging Facilities

A few RC systems—including Splash 1 and 2 [26], Wildforce [21,160], Pamette [366], and SLAAC1-V [39]—have provided observability through the ability to read back the configuration bitstreams of the Xilinx FPGAs. As mentioned above, this data provides user state information that can be used for debugging. Reading this data back requires hardware support for performing the configuration reads and software support in the form of at least an API for accessing the data. Providing an additional level of help in debugging, Splash 2 and Wildforce provided tools for automatically extracting symbols related to flip-flop state from a user's design and associating them with locations in the readback bitstream. As another debugging aid, the run-time systems of

Splash 2 and Pamette also provided support for symbolically retrieving the values of state elements and signals from configuration data readback. At the cost of some additional logic and design speed, several of these systems also provided direct access to the on-board memories from the host to ease application debugging and to allow easy data communication between the host and the application.

A large majority of RC systems provide little or no support for controllability. As exceptions, the SLAAC1-V board (and a few others) did support Virtex partial reconfiguration, which provides controllability of RAM elements only, and various boards based on the Xilinx XC6200 FPGAs provided complete controllability of the state elements (flip-flops). The support in both cases is limited to low-level access to the configuration mechanisms of the FPGAs. No high-level tools were included with the boards to provide users with a way of using partial reconfiguration for debugging controllability.

Unlike more recent systems, most of the above RC boards and systems provided some form of execution control. Typical execution controllability features for the boards mentioned above included single-stepping, multi-stepping, starting, and stopping clocks. Single-stepping means that a system can be executed one cycle at a time where, with multi-stepping, the system can be executed for a specific number of clock cycles before stopping. SLAAC1-V also included the ability to halt the system clock based on user signals. The run-time environments bundled with Splash 2, Pamette, and the SLAAC board family provided the designer with direct control over the clock so it can be started, stopped, or stepped.

SoftProbe

Beyond single-stepping and other useful debugging features, National Semiconductor's CLAy SDK board [307] provided a novel method for providing design observability. National's SoftProbe run-time environment allowed designers to sample the outputs of internal logic cells by routing signals from each of these outputs to a specific I/O pin on the FPGA so each value could be sampled individually. Before a debugging session (i.e., off line), a bitstream was created for each output signal to be sampled. Each bitstream would partially configure the CLAy FPGA to route the logic cell output at run-time to the predefined I/O pin. Since the partial reconfiguration used in this situation could change the operation of the circuit, the clock for the FPGA was stopped during the sampling operation so that no corrupt data was latched into any flip-flops. Further, the partial bitstreams could not route through any cells that were being used to create latches by cross-coupling gates. If signals were routed through these cells, the state held in the latch would be lost. To restore the design to its original form, the FPGA was fully reconfigured with its original bitstream.

Teramac

The HP Teramac project [18] added several unique observability and controllability features to their hardware and software system. Considering the entire system—from the FPGAs to the run-time execution environment—was designed at HP Labs, the designers were able to create a system that had unprecedented levels of observability and controllability. In the HP PLASMA FPGA, [19] the LUT outputs, register contents, and the system's memories were all available to the designer to observe through the *tmac* [208] run-time and debugging system. The *tmac* system with some additional tools allowed users to automatically correlate logical design signals and components with their sampled state across a system with as many as 16 boards.

Going beyond simple hardware observability, the *tmac* run-time environment could also reconstruct logical signal values that do not have physical counterparts. Often when using LUT-based logic, a group of signals and gates will be mapped to a LUT and not be observable. Teramac's run-time system took the known state of signals and partially simulated the logical (as opposed to the physical) design to reconstruct these signals which have been folded into LUT logic, thus, providing the designer with a significant increase in the design's observability. This concept would later inspire the hardware debugging framework for JHDL [182, 223].

With these observability capabilities, *tmac* also provided the designer with the ability to create custom graphical windows for interacting with the design and displaying design state as well as the ability to filter the state information to select interesting data. For instance, this capability was used to graphically display the progress of a simulated annealing partitioning algorithm as it executed. In another case, the HP designers were able to use the filtering capability to display disassembled versions of programs as they executed on a processor implemented on Teramac [66]; this was, of course, much easier than observing and deciphering the machine code that the processor executed.

Unlike the run-time software provided with other RC systems, the *tmac* run-time system for Teramac allowed the user to modify arbitrary flip-flop values at run-time. Since no primary inputs to the system actually existed, this controllability technique was necessary to provide inputs to the system. As mentioned above, the designer could even configure the run-time environment to provide a graphical interface for controlling state values while executing and debugging.

In addition to these observability and controllability features, Teramac provided the designer with the ability to specify a hardware breakpoint signal that, when asserted, would halt system execution. This differs from other RC systems which simply provided the API for such a feature.

InnerView

The InnerView Hardware Debugger [196] was part of the Virtual Wires [32] logic emulation project. The system required the user to instance their designs into a special design wrapper to use the debugging facility. As in Splash 2, the InnerView debugger used the configuration readback capability of the system's Xilinx FPGAs to sample the state of flip-flops and then display their values. The designer could define conditions within the hardware that would trigger readback operations. To preclude having to re-build the hardware for each trigger condition, all of the trigger functions were inserted into the hardware at one time and the system provided a way of masking the triggering of readback on a trigger by trigger basis.

Hardware Debugging Tools

For the last several years, both Altera and Xilinx have provided some hardware debugging tools that are independent of the system being used, assuming only that the system provides access to the FPGAs' JTAG interfaces. Specifically, Altera and Xilinx have been offering embedded logic analyzer systems—called SignalTap [12] and ChipScope [423], respectively—to provide FPGA designers with the ability to sample the state of their circuits under certain conditions. User designs are instrumented with these logic analyzers during either design entry or the low-level tool flow. Later versions of these tools—SignalTap II and ChipScope Pro—support incremental changes in the signals to be monitored, requiring only a partial recompilation of the design (SignalTap II) or a simple reroute of some signals in an already placed and routed design (ChipScope Pro). ChipScope Pro also provides a bus analyzer for on-chip PowerPCs or embedded Microblaze processors to help a designer debug processor bus related issues. To minimize the impact on designs, the FPGAs' JTAG ports are used to communicate with and control these analyzers. Though useful, these tools are not generally integrated into RC systems' development systems or run-time environments, requiring additional burdens to use as well as some additional hardware expertise.

Reconfigurable Computing Application Development Systems

Some of the larger challenges for debugging RC applications stem from the fact that the application development environment and the application execution environments are generally different and separate, complicating the design verification and debugging processes. Designers generally use a hardware description language (HDL) or schematic capture to describe RC applications and then use existing commercial simulators and synthesis tools for validating and implementing the designs, respectively. Then, when the hardware is ready for execution, the designer controls the RC system using a program and device drivers which have been developed separately from the hardware description. These software components are ordinarily created using common software languages (C, C++, etc.) and tools (compilers, linkers, assemblers, etc.). The RC development environment usually comes with libraries that provide the developer with an application programmer's interface (API) for interacting with the system.

This separation between the hardware design and execution environments results in several problems. First, the effort of validating an RC application in simulation may not significantly contribute to the effort required for validating the application in hardware—separate validation techniques and software must be written to perform much the same function. Second, controlling the simulation model and the actual hardware identically may be hard due to the differences in the simulation and hardware execution environments.

This suggests that there might be some benefit in tying the simulation and hardware execution environments together within a single hardware debugger for RC systems. This section will describe a few research projects that have attempted to tie these environments together, specifically, JHDL [181,182] and Sea Cucumber [204,205]. We also briefly describe the debugging capabilities of JBits—a very low-level application development tool—and a commercial debugging tool that has been developed to integrate with HDL synthesis.

JHDL

Though JHDL was originally conceived as a design tool for run-time reconfigurable systems [42], it later developed into a flexible, structural design tool enabling designers to create high-performance, tightly crafted hardware such as parameterized modules [221]. As a part of the design concept, JHDL was organized so that it could be used for the development and simulation of RC applications as well as an execution environment for these same applications.

Debugging support for applications was specifically included in JHDL's design [223]. With regards to debugging, the goal was to provide a system that allowed users to seamlessly use either simulation or hardware execution for debugging applications. All of the same circuit and circuit state viewers could be used in either the simulation or hardware execution modes. This ability to use both simulation and hardware for design validation was powerful since it provided the designer the ability to do validation before design implementation or compile the hardware and use the hardware itself for debugging. By using hardware directly for debugging, the designer has the opportunity to perform design validation on the final application rather than a simulation model, which, on occasion, may not completely match the actual hardware. Further, design errors that occur deep into an application's execution can be found more quickly with hardware debugging support—hardware execution is, of course, many orders of magnitude faster than simulating the hardware's execution.

One key to providing seamless operation between hardware execution and simulation was the JHDL logic simulation kernel. JHDL employed a levelized logic simulator [176] that statically scheduled the order in which combinational logic functions would be evaluated based on input/output dependencies. With this approach, the simulation kernel can simply evaluate all sequential elements and primary inputs first and then propagate the values of combinational elements based on the static evaluation schedule, as illustrated in Figure 4.12.

To support the seamless execution and debugging of an application in hardware, the simulation kernel was extended for "hardware mode" as shown in Figure 4.13. In this version of the kernel, hardware is executed for a cycle (or several cycles) at the beginning of a kernel iteration. Then, the complete hardware state is read—including flip-flop and on-chip RAMs—and this state is inserted into the simulation model. Next, the synchronous elements that were not updated from hardware are evaluated (or clocked) and, finally, the combinational logic elements are evaluated based on the levelized schedule.

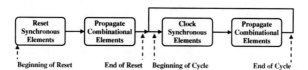

Fig. 4.12. JHDL Logic Simulation Kernel

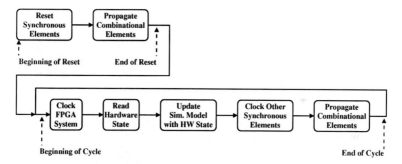

Fig. 4.13. JHDL Logic Simulation/Hardware Kernel

In contrast to a levelized logic simulator, an event-based simulator [176] uses a more dynamic evaluation approach by evaluating circuit functions and signal values as events happen. Though event-based simulation can be more efficient since it only evaluates what it needs to, integration of event-based simulation with hardware execution is more difficult than integration with levelized simulation.

The other key to seamless simulation and hardware execution was the support in JHDL for mapping the state from the physical hardware to the logical view of the user's design. As illustrated in Figure 4.14, a significant amount of information and effort was required to perform a complete mapping. To begin, with JHDL provided a list of all state elements to find, illustrated

as the .rbsym file in the figure. The information provided by the Xilinx design
implementation tools were parsed to identify the physical locations of the state
elements and the locations of their state in the readback bitstream. Finally,
the technology mapping report file (.mrp file) was used to identify how signal
names were modified during the technology mapping phase to resolve how the
address signals were connected to LUT RAM address inputs. The product
of the process, .rbentry file, was effectively the symbol table used during
hardware execution for relating hardware design state with the logical circuit
representation used by JHDL.

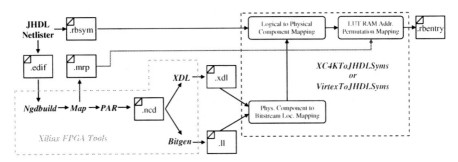

Fig. 4.14. Logical to Physical Design State Mapping for JHDL

Overall, this approach accomplishes what *tmac* did for Teramac's execution environment in the sense that the values of signals and logic elements
that are not found in the physical design (i.e., those packed into LUTs) can
still be observed through a combination of hardware execution and simulation.
Unlike *tmac*, JHDL was designed to work with commercial Xilinx FPGAs and
software as well as FPGA boards from different sources, so the JHDL system
had to be able to integrate its designs with external hardware API support
and the Xilinx design tools to provide this debugging support.

As more experimental additions to JHDL, several other debugging features
were added. First, as described in [414, 415], an automated method was developed to create scan chains of all design state elements in a user FPGA design.
Similar in concept to scan chains used in JTAG or other design for testability
techniques, these scan chains could be used to both sample the state of a user
design (observability) as well as to force the design to a certain state (controllability). Though JHDL's standard method of sampling a circuit's state
was using the FPGAs' configuration readback capability, JHDL's hardware
execution and debugging system readily accommodated the transfer of state
to and from designs' scan chains.

Additional work reported in [181, 182] leveraged bitstream manipulation
techniques to provide observability and controllability. For example, JHDL
was extended so that it easily allowed designers to include embedded logic
analyzers in designs as well as to quickly modify designs to connect signals to

the logic analyzers. Users could use the schematic viewer to select the signals
to trace and to define the triggers for state capture. Once these were defined,
JBits would then be used to modify designs' bitstreams to make these connections to the logic analyzers for the current debugging session. The results
captured by the embedded logic analyzers were available within the JHDL
environment in the form of a trace buffer viewer. Instead of using JTAG to
communicate with the logic analyzers, configuration readback was used to
sample the values of the on-chip trace buffers. The benefit of modifying the
bitstreams directly when using these logic analyzers was that it performed the
modifications 5 to almost 20 times faster than by using other more conventional techniques.

Sea Cucumber

Taking a higher-level approach, the debugger [205] for the Sea Cucumber (SC)
synthesizing compiler [397] provides for a more software-like debugging experience than JHDL since the design is captured as a collection of Java threads
that communicate via Communicating Sequential Process (CSP) channels. As
with JHDL, the debugger can be used to debug the design in either simulation
or hardware-execution mode—a significant improvement over the debugging
capabilities of other high-level tools such as SystemC or Handel-C where only
the software description can be debugged and not the final circuit.

The feature set desired and achieved with the SC debugger was:

1. to allow single stepping of program execution (execution control),
2. to allow arbitrary break points in the code (execution control),
3. to allow variables in the source to be set to arbitrary values (controllability),
4. to display the current execution points while execution was paused (observability), and
5. to allow program variables to be observed (observability).

To develop a debugger with this feature set, several significant challenges
had to be overcome, including the unconstrained nature of the hardware as
compared to microprocessors, the presence of conventional and VLIW-style
compiler optimizations during synthesis that modify how the code executes
and the variable values, and the mapping of hardware state and execution to
SC design descriptions. To aid in the latter, JHDL was actually used as the
lower-level HDL to which SC compiled so that JHDL's facility for mapping
the actual hardware state to the structural description of the hardware and
providing the hardware state during hardware execution could be used. Of
course, the SC debugger still had the difficult task of mapping the structural
hardware state to the Java programs from which the hardware was compiled. The project illustrated that hardware debugging support was possible
for high-level synthesis systems even in the presence of significant compiler
optimizations.

JBits

Beyond the novel bitstream manipulation capabilities mentioned earlier, JBits also provided a few utilities for hardware debugging. First, as with several systems already mentioned, BoardScope could be used with the configuration readback capability of Xilinx FPGAs to sample a chip's state. The instantaneous sampled state values could be displayed graphically for an entire chip or the state values could be displayed over time using a waveform viewer. If a design was implemented using certain JBits Core APIs, the cores themselves could interpret and display their own state.

Using a hardware interface API called the Xilinx Hardware Interface (XH-WIF), BoardScope and a scripting tool called DDTScript could control RC and other FPGA-based systems. XHWIF provided the ability to reset the system, set the board clock frequency, start and stop the clock, step the clock a given number of cycles, configure FPGAs, read back FPGA configuration data, and read and write the contents of FPGA system memories. Clearly, the XHWIF interface provides little more than a software interface between the JBits environment and existing board capabilities and APIs, thus, the board and supporting board API must provide this functionality in some way for XHWIF to be useful. BoardScope also provided a novel but limited facility for simulating the execution of hardware designs using only FPGA configuration bitstreams.

Identify

To conclude the discussion of RC application debugging, Synplicity—a company producing tools for the synthesis of VHDL/Verilog FPGA designs—has developed a tool called Identify [385] that supports HDL debugging using hardware. Using either a built-in JTAG interface or a user defined JTAG interface, the Identify system can communicate with a user design instrumented for debugging. The designer can set breakpoints within the HDL code based on certain branching events, a certain sequence of events, or specific signal conditions—providing an intuitive method for generating debugging triggers for the HDL designer. It can handle multiple clock domains simultaneously and can handle symbolic data instead of just bit-level data. Synplicity has made an effort to reduce the impact of instrumentation in terms of design area and speed.

4.5.3 Challenges for RC Application Debugging

Past tools and systems have demonstrated what is possible for debugging reconfigurable computing applications and what can be improved, but, despite these advances, the most advanced tools available today for debugging tend to be trial and error, conventional logic analyzers, and tools like Identify, SignalTap, and ChipScope. Though each of these tools can be very useful, they illustrate the lack of real debugging support in most RC systems, which has a significant effect on the productivity for developing validated, complex reconfigurable computing applications.

Part of the reason for this is that many treat the development of RC applications for FPGAs and similar devices to be essentially the same as the development of ASIC hardware. However, the two development cycles are quite different. In the case of ASICs, enormous amounts of time are required to simulate the designs because it is extremely costly to fix problems with the hardware once it has been fabricated—it requires $1 million dollars or more for a new mask set alone for a VLSI design. These economics do not exist, of course, with reconfigurable computing since the hardware is often available for use and the cost of changing the design is comparatively small. Of course, with design recompilations taking hours for very large designs, simulation is still important for RC application validation. On the other hand, the simulation of thousands of cycles can take as much as a day or more for large complicated designs, so the recompilation cycle for hardware is not always unreasonable when justifying debugging using the actual hardware.

Another difficulty is that of economics for RC debugging systems. Unlike the developers of conventional software debuggers that can be retargeted to many systems somewhat easily due to the fact that the target microprocessor (and the related system) is fixed, the developers of debuggers for reconfigurable computing applications are faced with having to support considerably smaller system volumes and the costs of developing the debugging support for each different RC system, currently, can be quite large. To help this situation, some standardization of debugging support across reconfigurable computing systems might encourage third-party debugging tools. Further, to generate the information needed to support hardware debugging using commercial EDA tools would also take the willingness and concerted effort by RTL synthesis, FPGA, and RC systems companies. As reconfigurable computing becomes more common, hopefully these technical and non-technical challenges can be addressed so better debugging support exists for reconfigurable computing applications.

4.6 Summary

Reconfigurable computing requires the description of algorithms, their expression into software and hardware circuits, and the debugging of the resulting software/hardware systems. RC languages range from C, Fortran, and Java variants, including parallel language variants, to graphical programming languages. Novel language features include directives to partition computation, to allocate variables to specific memories, parallel processing directives, directives to pipeline lops. Compilers for RC languages face formidable challenges in generating efficient hardware circuits. In contrast to fixed instruction microprocessors, micro-architecture design decisions must be resolved during compilation. Once an application-specific architecture has been designed, the inherent flexibility of FPGAs make the tasks of synthesis, mapping, placement, and routing difficult and time consuming. Finally, debugging reconfigurable systems imposes unique challenges to gain visibility into the hardware as well as execution control during debugging.

5

Digital Signal Processing Applications

This chapter will discuss digital signal processing as an application domain for reconfigurable computing (RC). To outline the chapter, we will first discuss what is meant by digital signal processing (DSP) generally and then describe why reconfigurable computing is well suited for DSP. The remaining sections describe common operations performed for DSP as well as some example DSP applications and their reconfigurable computing implementations. Note that digital image processing, a subset of DSP, is covered in Chapter 6.

5.1 What is Digital Signal Processing?

As the growing number of texts [293, 314, 332] on the subject suggest, digital signal processing involves the representation of signals digitally as sequences of numbers or symbols and the processing of these sequences to extract information from the signals or to synthesize signals with desirable properties either as completely new signals or from existing signals. Figure 5.1 illustrates a typical DSP system that takes analog signals as inputs and produces analog signals as outputs.

Fig. 5.1. Simplified Digital Signal Processing System (based on [293])

For many decades, analog techniques and circuitry that deal with signals in their natural, continuous forms were the methods of choice when processing signals. As computers and digital hardware have advanced, the use of digital signal processing techniques has become more feasible and very common place, being used for recording, storing, and reproducing both sound and video signals digitally; enabling wireless communications; enabling advanced medical imaging and other diagnostic helps; and a variety of other applications. This transition from analog to more digital techniques has been driven by the many advantages of DSP, including [293]:

- the increased immunity to changes in external parameters such as age and temperature as well as to variations in circuit components;
- the ease of reproducing DSP systems;
- the flexibility in precision through changing word lengths and/or numeric representation (e.g., fixed point vs. floating point);
- the ability to use a single processing element to process multiple incoming signals through multiplexing;
- the ease with which digital approaches can adjust their processing parameters, such as with adaptive signal processing; and
- the ideal characteristics that are only possible through digital techniques (e.g., exact linear phase, multi-rate processing, the lack of loading effects when cascaded, the ease of signal storage and reproduction, very low frequency processing, etc.).

The disadvantages of using digital techniques over analog ones often include increased system complexity, power consumption, and frequency range limitations. Regarding system complexity, DSP requires that signals be converted between analog and digital forms using sample-and-hold circuits, analog-to-digital converters (ADCs), digital-to-analog converters (DACs), and analog filtering. As well, DSP can employ relatively complex digital devices for processing. By contrast, analog techniques can get by with passive components and significantly fewer active components. This complexity issue also directly affects power consumption—digital processing tends to require more power. With regards to frequency range limitations, analog hardware will naturally be able to work with higher frequency signals than is possible with DSP hardware due to the limitations of performing analog to digital conversion. For many applications, the advantages of DSP far outweigh these disadvantages.

Some common operations performed on signals using digital or analog techniques include:

- elementary time-domain operations (amplification, attenuation, integration, differentiation, addition of signals, multiplication of signals, etc.),
- filtering,
- transforms,
- convolution,
- modulation and demodulation,
- multiplexing and demultiplexing, and
- signal generation.

Among other possibilities, combinations of these operations can be used to remove noise from signals, prepare signals for wireless transmission, extract signals from wireless transmissions, or select signals of interest from a larger collection of signals.

5.2 Why Use Reconfigurable Computing for DSP?

This section will describe why reconfigurable hardware computing can be so well suited for DSP. Following that discussion, we will briefly describe how reconfigurable hardware and computing approaches compare with other DSP implementation technologies.

5.2.1 Reconfigurable Computing's Suitability for DSP

Numerous studies in implementing DSP applications using reconfigurable computing hardware have demonstrated that reconfigurable computing can be well suited for digital signal processing [79, 326, 389]. The effectiveness of reconfigurable computing for DSP is mainly due to the parallelism that can be exploited in DSP applications. The factors contributing to this parallelism are described below.

First of all, many DSP applications inherently have significant amounts of both fine-grained as well as coarse-grained parallelism. As a simple example, consider how a finite-impulse response (FIR) filter can be implemented in an hardware. A FIR filter implements the following equation:

$$y(k) = \sum_{n=0}^{N-1} a(n)x(k-n) \tag{5.1}$$

A signal flow graph representation of the filter is depicted in Figure 5.2. Assuming that the filter has no symmetry, a microprocessor-based solution would perform the computation's N multiplications and $N-1$ additions sequentially. In reconfigurable hardware, FIR filters can be implemented using pipelining so that N multiplications and $N-1$ additions are performed each cycle with a result provided every cycle. Figure 5.3 illustrates an equivalent FIR filter that has been highly pipelined.

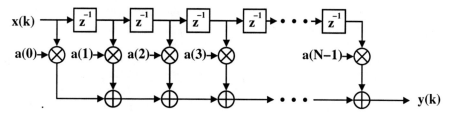

Fig. 5.2. Simple FIR Implementation

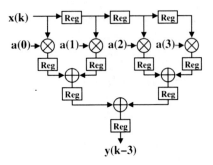

Fig. 5.3. Highly Pipelined FIR Implementation (4 Taps)

Of course, very few DSP applications consist of a single filter. Most DSP applications require several operations such as FIR filters, transforms, etc. to process each incoming data stream, providing the potential to exploit coarse-grained parallelism as well. For example, Xilinx's digital down converter [425], illustrated in Figure 5.4, illustrates multiple coarse-grained operations (signal multiplication and many filters), each having fine-grained parallelism internally.

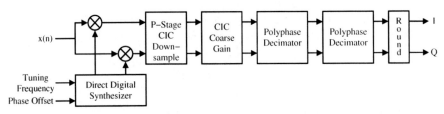

Fig. 5.4. Xilinx Coregen Digital Down Converter (V. 1.0)

As illustrated with the FIR filter, many DSP functions have relatively regular schedules for their operations, allowing the hardware to be customized to extract a significant amount of parallelism. This regularity also reduces the control logic burden for applications. By contrast, applications requiring significant amounts of nested decision logic (e.g., multiple levels of nested if...then constructs) tend to limit the parallelism that is possible since the computations performed depend heavily on the path through the decision logic—a path that may not be easy to predict.

A factor that helps the exploitation of parallelism in DSP applications is the small word widths that many DSP applications naturally use for their data. These word widths are often the result of the data precision provided by the analog-to-digital converters—many produce 8- to 12-bit data words. When compared with floating-point implementations, these smaller word widths require smaller arithmetic units and less routing, resulting in higher operating clock rates and allowing many units to be instanced on the same chip. Of course, internal data widths in a DSP application will grow past the input precision and a significant amount of effort may be necessary to balance word widths with the precision necessary for an application. If this balance can be maintained, a single reconfigurable device can perform a significant number of operations.

Another issue that improves the parallelism that can be exploited by reconfigurable computing is the fact that DSP applications often use fixed coefficients or constants throughout their computations. By "folding" the constants directly into the hardware, i.e., customizing the hardware for a given constant, the area and speed of operations can be significantly improved. For instance, a signed 16-bit by 16-bit multiplier implemented using Virtex-II slices requires about 184 slices (based on Xilinx ISE 6.3 Coregen's Multiplier 7.0 core generator). If a 16-bit by 16-bit multiplier is optimized assuming one of the inputs is forced to be -23131 (the signed decimal value of the hex constant A5A5), the multiplier's size reduces to only 90 slices (again, based on the same core generator)—less than half the size. Of course the reduction can be more dramatic, such as the multiplication by a power of two, which results in a simple left shift of the value (essentially some simple wiring and the addition of some zeros in the least significant bits).

Reconfigurable computing's ability to supply both flexible and significant memory bandwidth also improves the possible parallelism that can be extracted in DSP applications. Referring again to the FIR example illustrated in Figure 5.3, note that the hardware is providing 15 different data values simultaneously each cycle, including $x(k-3)$ to $x(k)$, $a(0)$ to $a(3)$, the partial results, and the $y(k-3)$ result. The storage for $x(k)$ and the a coefficients are not shown explicitly, but they could easily be supplied by flip-flops or on-chip RAM. If the on-chip memory bandwidth only allowed a few values to be manipulated at a time, only a single operation could be performed at a time instead of all seven simultaneously.

When compared with a sequential microprocessor, which generally manipulates only two input values and one output value per instruction and may have a few instructions processed per cycle, this internal register and memory bandwidth can be significant, especially considering entire applications. For example, as mentioned in the Chapter 2, the Altera EP2S180 can theoretically supply an aggregate memory bandwidth of over 30 Gb/s through its 3414 ports to its 1707 on-chip SRAMs, not including the contribution of its almost 180,000 on-chip flip-flops.

Further, several reconfigurable computing systems provide ten or more ports to external SRAM as well. These ports to deeper external memories can provide the ability to process multiple blocks of data simultaneously, providing the opportunity to exploit additional parallelism through partitioning a single data set across multiple DSP computation engines or allowing the processing of multiple input streams or data sets. Clearly, the ability to customize an application's memory hierarchy and the availability of on- and off-chip memory bandwidth through many independent memory ports contributes significantly to application performance and the parallelism that can be exploited by reconfigurable computing.

Similar to the issue of memory bandwidth, input/output (I/O) bandwidth can also have a significant impact on DSP performance. Unlike with microprocessor-style approaches that must retrieve the data from a system peripheral, the output of one or more analog-to-digital converters (ADCs) can be driven directly into reconfigurable computing hardware (such as an FPGA), significantly reducing the overhead for providing input data. Further, with the availability of as many as 1000 or more user I/O blocks and with the addition of multi-gigabit serial transceivers, the high I/O bandwidth available among reconfigurable hardware devices also provides a more scalable way of using multiple devices for a given application than is possible with conventional microprocessors. In addition to the bandwidth possible, the flexibility of this I/O can allow designers to customize the I/O to the needs of a given application, allowing for the better use of I/O bandwidth.

Based on the above discussion, DSP and reconfigurable computing can be well matched due to the available parallelism and the efficiency of custom reconfigurable computing implementations. Though many reconfigurable computers only operate at tens to a few hundred megahertz, given the above potential for exploiting parallelism, it should be no surprise that reconfigurable computing implementations of DSP applications can outperform microprocessor and programmable digital signal processors despite operating at 1/10th or less of the clock rate.

5.2.2 Comparing DSP Implementation Technologies

Once the decision has been made to perform the signal processing digitally, many options exist for the type of hardware used for the processing. Some of these options include general-purpose processors (GPPs), microcontrollers, programmable digital signal processors (PDSPs), reconfigurable logic or computing hardware, and application-specific integrated circuits (ASICs). Each of these options has its place in the spectrum of DSP systems and applications. Table 5.1 provides a qualitative comparison of several of this implementation technologies.

In this design space, general-purpose processors provide relatively low performance due to the bottlenecks of their von-Neumann-style architectures (i.e., shared data and instruction memory). Because GPPs and GPP systems are produced in volume and high-level programming is well supported, system-design and per-chip costs are relatively low for this technology. GPPs and their systems can require considerable amounts of power due to the support logic required and the relatively low amount of throughput provided by the systems. The greatest attraction besides cost to use GPPs is their flexibility. GPPs can easily be programmed and reprogrammed to handle new tasks, as

Technology	Performance	System Design Costs	Cost Per Chip	Power	Flexibility	Memory Bandwidth	I/O Bandwidth
General-Purpose Processor	Low	Low	Low-Medium	High	High	Low	Low
Microcontroller	Low	Low	Low	Medium	High	Low	Low
Programmable DSP	Medium	Medium	Low-Med.	Medium	Medium	Medium	Low
ASIC	High	High	Low	Low	Low	High	High
Reconfigurable Hardware	Medium-High	Medium	Medium-High	Medium-High	High	High	High

Table 5.1. Qualitative Comparison of DSP Implementation Technologies

necessary. Compared with what is possible with custom hardware, the memory bandwidth and I/O bandwidth possible with GPPs is comparatively low. Frequently, GPPs are used for DSP when they are already available in the system (such as with desktop PCs) and the application does not need high performance.

Microcontrollers, a close relative to the GPPs, are effectively lower performance microprocessors that are used more for control applications rather than data processing. They frequently have support hardware for doing I/O and sometimes special instructions for performing DSP functions. For the most part, their strengths and weaknesses are similar to those of GPPs, except they generally require less power and provide less performance.

Programmable digital signal processors (PDSPs) are a implementation technology that has become very popular. Their heritage is with GPPs, considering the first PDSPs were basically general-purpose processors with a few architectural adjustments for performing filtering and transforms. Later generations have moved further away from GPPs, using a Harvard architecture (separate instruction and data busses) to improve memory bandwidth for computation, adding features for low-overhead looping, providing hardware address generators, and even adopting very-long instruction word (VLIW) architectures. As a result, modern PDSPs can easily overlap data fetches, computation, address generation, and loop control.

With regards to the comparison of Table 5.1, PDSPs provide more performance than GPPs and microcontrollers at lower power, but the cost of implementing a DSP algorithm is higher due to the added difficulty of programming them—a difficulty related to the specific ways their special architectural features must be used for peak performance. High-level programming tools are available, but they are not as effective as those for GPPs due to the DSP's special architectural features. This issue also affects their flexibility—they can be retargeted to perform different applications, but not all applications map well to PDSPs. Because of their performance and power characteristics, PDSPs are now found in many lower-power applications such as compact-disc players and other digital music players, conventional and DSL modems, wireless telephones, and even hard-disk drives. High-end PDSPs are, of course, available for higher performance at the expense of more power and cost.

Probably the most ideal technology for performing DSP in terms of performance, power, and both memory and I/O bandwidth is the ASIC since it can be customized to the application at hand. The biggest problems are the costs of creating custom silicon solutions ($> \$1$ million per mask set for state-of-the-art CMOS processes and requiring months to a year or more to design, fabricate, and validate) and the lack of flexibility. ASICs can be designed with some flexibility in terms of handling different sets of coefficients or slightly different computations, but they tend to be considerably less flexible than all other solutions for implementing completely different applications—once the design is fixed in silicon, it is not going to change. ASICs are generally reserved for very high volume applications (for example, MPEG (-1, -2, or -4) encoding or decoding for wireless telephones) or for mission critical applications where the performance and other constraints are paramount (e.g., military applications). High-volume mixed-signal ASICs that integrate ADCs and DACs directly with the DSP hardware are also common for consumer applications.

In comparison with these other technologies, reconfigurable hardware provides a performance-flexibility compromise between PDSPs and ASICs. Their performance can often be better the PDSPs while being one-tenth the performance of an ASIC implemented in the same CMOS technology. If FPGAs are used, the cost of system development lies between that of PDSPs and ASICs since FPGA design is hardware design, generally. Reconfigurable hardware requires more power than ASICs and, sometimes, even PDSPs (assuming the PDSP performance is adequate). FPGAs tend to be more expensive than PDSPs or high-volume ASICs on a per chip basis, though, some higher-volume FPGAs are now available for tens of dollars. High-end FPGAs can cost thousands of dollars even as high as $10,000 or so per FPGA. FPGAs provide a large amount of flexibility since they can be adapted effectively to wide variety of new applications. Further, since the memory and I/O interfaces can be customized, the memory and I/O bandwidth available to FPGAs can be significantly higher than with other programmable solutions, though, not necessarily as high as ASICs. For now, reconfigurable computing solutions tend to be excellent for high-performance, low-to-mid-volume applications where system power constraints can be met. If, in addition, reprogrammability and flexibility are also needed, the reconfigurable hardware solution is very well suited.

The above analysis is similar to that found in [389], but, over the last several years, a very interesting trend has developed. With the continually increasing costs of developing new ASICs, fewer and fewer ASIC solutions are available except for the very high volume applications. Further, unlike during the 1980's and 1990's, FPGAs now tend to lead other CMOS ICs when it comes to the CMOS fabrication technology used. As a result, the highest performance implementations available for some DSP applications may actually be FPGA-based as FPGA technology continues to progress in terms of area and speed.

As an example of this recent trend, we did a study in 2004 of complex 4096-point Fast-Fourier Transforms (FFTs). Table 5.2 provides a performance comparison among the several commercial ASIC and FPGA implementations available. As the data illustrates, the FPGA implementations were considerably faster than the other fixed-point FFTs. Also, the ASICs, which performed floating-point or block floating-point operations, were slower than the Virtex-II- and Virtex-II-Pro-based floating-point FFTs[1]. Note that the fastest ASIC

---

[1] Note that two of the numbers are actually for 2048-point transforms and, thus, a 4096-point transform by these chips (if they were possible) might take about twice as long. Thus, the FPGA implementation is faster for the longer transform.

(the Eonic) employed a 180-nm CMOS process while the FPGAs employed newer CMOS processes.

Implementation Technology	CMOS Tech.	FFT Size	Input Data Format	Clock Rate	Time per FFT	Power
Dillon Eng. (Virtex-II Pro)	130 nm	4096	2x18-bit, fixed pt.	200 MHz	3.84 μs [125]	< 4 W [125]
Dillon Eng. (Virtex-II)	150 nm	4096	2x18-bit, fixed pt.	160 MHz	4.8 μs [125]	< 4 W [125]
Pentek Virtex-II (4954-404, 4 overlapped FFTs)	150 nm	4096	2x16-bit, fixed pt.	160 MHz	6.4 μs [376]	5 W [375]
Pentek Virtex-II (4954-403, no overlap)	150 nm	4096	2x16-bit, fixed pt.	140 MHz	29.3 μs [325]	1.5 W [375]
TI 320C64XX (VLIW DSP)	130 nm	4096	2x16-bit, fixed pt.	600 MHz	48 μs [376]	1.5 W (typical) [209]
Motorola MPC7455 (Altivec, VSIPL)	180 nm	4096	2x16-bit, fixed pt.	1 GHz	53 μs [376]	8 W (typical) [300]
Dillon Eng. (Virtex-II Pro)	130 nm	4096	2x32-bit, floating pt.	200 MHz	30.7 μs [125]	< 4 W [125]
Dillon Eng. (Virtex-II)	150 nm	4096	2x32-bit, floating pt.	160 MHz	38.4 μs [125]	< 4 W [125]
Eonic PowerFFT (ASIC)	180 nm	4096	2x32-bit, floating pt.	128 MHz	48 μs [138]	< 2 W [138]
Radix RaCE FFT (ASIC, 2 Chips)	N/A	2048	2x18-bit, bl. floating pt.	84 MHz	24.4 μs [336]	< 3.5 W [336]
CRI Pathfinder 2 (ASIC)	250 nm	2048	2x32-bit floating pt.	120 MHz	31 μs [124]	< 2.5 W (estimated)

Table 5.2. Comparison of Commercial ASIC and FPGA FFT Implementations

Again, this trend does not necessarily hold for very high-volume DSP functions implemented as ASICs, but lower volume ASICs have started to be taken over by FPGA-based implementations. Also, as FPGAs start to integrate fixed DSP hardware into their architectures (see Section 2.1.2), their roles in high-performance DSP may continue to expand.

5.3 DSP Application Building Blocks

Considering the popularity of performing DSP using reconfigurable hardware, a complete review of all the significant work implementing efficient DSP structures in FPGAs and reconfigurable hardware is beyond the scope of this chapter. We refer the reader to [326, 389] for some overviews of early and more recent work. Instead, this section will briefly describe a few significant, basic building blocks and techniques frequently used in reconfigurable computing for DSP, citing a few examples of related work.

5.3.1 Basic Operations and Elements

The most basic components of most DSP applications are arithmetic operations and memory elements. The most common arithmetic operations include addition, subtraction, and multiplication. Memory elements play a complementary role as temporary storage of results between operations, such as delay registers and "scratch-pad" memories. In fact, one common DSP building block, the multiply-accumulate (MAC) unit, is a combination of multiplication, addition, and memory, as illustrated in in Figure 5.5. This section describes a few methods that have been used in FPGAs to provide efficient arithmetic and memory structures for DSP, considering how frequently they are used.

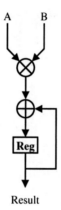

Fig. 5.5. Multiply-Accumulate Unit for DSP

Efficient Arithmetic

Many of the hardware-efficient techniques developed and used for arithmetic in VLSI and other hardware implementation technologies over the years can be applied to reconfigurable hardware such as FPGAs. The book by Koren ([243]) is a good reference for some of these techniques as well as for the fundamentals of computer arithmetic more generally. Below, we will briefly note a few of the many arithmetic techniques that have been utilized for efficient FPGA implementations of arithmetic for DSP applications.

Bit- and Digit-Serial Arithmetic

As described in [111], [326] and [389], bit-serial arithmetic has been used for signal processing hardware for some time, especially when latency is not as crucial as circuit area and I/O utilization. Because bit-serial arithmetic trades computation latency for design area, the resulting circuits are very small, need very few I/O pins, and often operate at high clock rates. Further, a arithmetic operation using N-bit operands generally requires at least N cycles to complete. Figure 5.6 provides an example of a bit-serial adder, which requires only a single 1-bit adder and a register.

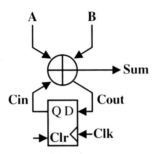

Fig. 5.6. Bit-Serial Adder

While common when FPGAs provided relatively few resources, bit-serial arithmetic is rarely used now that FPGAs provide significantly more resources. This is partly due to the complexity of correctly controlling such circuits and partly due to the fact that most design tools do not directly support this type of arithmetic. As an example of the control that is necessary, the flip-flop in Figure 5.6 needs to be cleared before the operands appear at the A and B inputs to get the proper result from the bit-serial adder. Further, the operands must be provided least-significant bit first and properly aligned. Despite these drawbacks, bit-serial techniques are still useful when area efficiency is required and the latency is not crucial.

Digit-serial arithmetic [199, 320], as the name suggests, breaks up the binary representation number into N-bit digits and these N-bit blocks are provided sequentially to the arithmetic hardware for computations. Again, this provides a trade-off between circuit area and computation latency, using additional clock cycles and relatively little hardware to perform a computation. While extending the possible design space for computer arithmetic and providing higher performance options than bit-serial implementations, digit-serial arithmetic has many of the same advantages and disadvantages of bit-serial approaches. As with bit-serial arithmetic, digital-serial arithmetic has been used in FPGA-based DSP designs (e.g., [254]) mainly due to early FPGAs' I/O and area constraints.

Distributed Arithmetic

Another hardware-efficient approach to arithmetic is distributed arithmetic (DA) [416]. This form of arithmetic received its name from the relatively unobvious way it implements arithmetic. Distributed arithmetic uses ROMs for looking up partial results that are then used in the computation. Since FPGAs utilize LUTs for logic, it is natural to apply distributed arithmetic to FPGA hardware. DA can be used with bit-serial, digit-serial, or fully parallel number representations and is used when the results or partial results of a computation can be pre-computed and stored in a ROM, such as when constants are used in a computation.

For example, Figure 5.7 illustrates a parallel distributed arithmetic implementation of multiplication by a constant value. The ROMs store the results of a 4-bit by 8-bit multiplication, where the 8-bit operand is constant. The 12-bit results of the two multiplications performed—the upper partial product (UPP) and lower partial product (LPP)—are then combined with an adder. As described in [326], this implementation required about 1/3 of the logic of a multiplier with two variable operands.

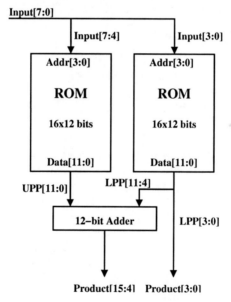

Fig. 5.7. 8-bit by 8-bit Parallel Distributed Arithmetic Multiplier [326]

CORDIC

While multiplication, addition, and subtraction are common to DSP algorithms, many algorithms require more complex functions. Again, enumerating all contributions to developing efficient hardware implementations for operations such as division and square roots would be voluminous. For this brief overview of DSP applications, we focus on one particular technique that has proven to be quite useful for DSP, specifically, the CORDIC algorithm.

The CORDIC (COordinate Rotation Digital Computer) algorithm is an area-efficient technique for computing some trigonometric, hyperbolic, and linear functions. First developed by Volder [405] for circular coordinate systems and later unified by Walther [411] to also include linear and hyperbolic coordinate systems, CORDIC computes the effects of rotating vectors in these coordinate systems using an iterative algorithm that simply uses additions, subtractions, and shifts with fixed-point data. Due to the algorithm's simplicity, CORDIC has been popular for VLSI and FPGA hardware for computing such functions as $sin\theta$, $cos\theta$, $tan^{-1}(y/x)$, polar-to-rectangular and rectangular-to-polar coordinate conversions, $sinh\theta$, $cosh\theta$, and several other functions in its unified form. It has formed the basis for such DSP operations as Discrete Cosine Transforms (DCTs) [71], Fast Fourier Transforms (FFTs) [121], sine-wave synthesis [122], computing eigenvalue decompositions [237], and solving singular value problems [216]. Good overviews of CORDIC can be found in [217] and [244]. For a survey of FPGA implementations of CORDIC as well as a brief description of the algorithm refer to [20].

Figure 5.8, which is based on [217], illustrates the circular rotation of a vector due to four CORDIC iterations. In the circular coordinate system, rotations follow a circle centered at the origin. In the linear coordinate system, the rotations follow a line parallel to the y axis. Likewise, the hyperbolic rotations follow a hyperbola.

As the figure illustrates, the desire to keep the arithmetic in each iteration simple has a few effects. First, the angles used are those that can be simply represented in fixed point. In fact, for the circular case, the angles are those angles ϕ satisfying the following relation: $tan(\phi) = \pm 2^{-i}$. Thus it takes several iterations to approximate most angles well. The second side effect is that the length of the resulting vector may have a gain factor. In the linear coordinate system, there is no built-in gain that affects vector lengths, while in the other two coordinate systems there are gains associated with each rotation iteration. For example, the simple circular coordinate algorithm results in vectors that have lengths approaching approximately 1.647 times their original size. Frequently, to keep the computations simple, these gains are factored into the design of algorithms and may or may not be compensated for during the computation. Another side effect of the gain in the circular coordinate system is the fact that the numeric representation for the quantities must also grow to accommodate this increase.

As described in [20], these side effects of the algorithm are tolerated because of the range of commonly used functions that can be computed by such an efficient structure. The iterative hardware requires three adder/subtracters, shifter logic, and, possibly, a ROM. The fully unrolled CORDIC unit requires three adder/subtracters per iteration and the shifts can simply be imple-

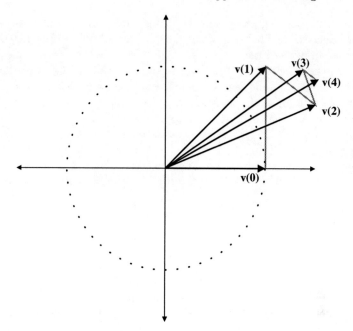

Fig. 5.8. Example of Four CORDIC Iterations in the Circular Coordinate System [217]

mented with wiring. A fully unrolled and pipelined CORDIC unit requires approximately as much area as an array multiplier.

Efficient Memory Structures

As FPGA architectures have evolved, the options for memory have also evolved to favor DSP applications. On the first FPGAs, the only memory available was in the form of flip-flops. While flip-flops are abundant in most reconfigurable architectures, many applications (especially image processing applications) can require additional memory for temporary data buffers. In more recent systems, flip-flops are used for pipeline registers and short-depth buffers when access to all values are needed.

As mentioned in Section 2.1.2, the LUTs in the logic blocks of Xilinx FP-GAs since the XC4000 series can be used as small RAMs. In fact, these LUT memories can easily be used as small shift registers or FIFOs to hold data. Since the Virtex series, Xilinx has made it easier to implement shift registers with these memories by providing the SRL16 functionality for the LUTs— a functionality that provides a 1-bit wide shift register with a dynamically changeable depth (from one to sixteen bits deep). Thus, for later Xilinx architectures, a 17-bit deep shift register can be easily constructed in half of a slice using a LUT in SRL16 mode and the associated flip-flop rather than using 17 individual flip-flops. LUT memories are thus appropriate for relatively shallow memories (e.g., <64 bits deep). LUT memories are also used to create efficient delay lines where all of the data in the delay line does not need to be accessible all of the time.

Also mentioned in Section 2.1.2 is the fact that FPGA now have larger embedded SRAM memories with depths greater than .5 Kb and programmable aspect ratios to allow the designer some flexibility in trading depth for width and vice versa to better match an application's needs. Many of these deeper memories also provide two memory ports for better concurrency. Each of the two ports can be driven by a different clock source and can operate with a different memory aspect ratio. These larger memories are useful for larger buffers and for high-speed, flexible FIFOs across clock domains.

5.3.2 Filtering

As discussed above, filtering is a common operation in digital signal processing and takes many forms. Filtering, as its name suggests, is used to remove unwanted components from signals while maintaining the components of the signal that are desired. This function is applied in many situations. For instance, there are decimating filters, which are used to reduce the effective sample rate of signals, and interpolating filters, which are used to increase the effective sample rate of signals. Below we will briefly discuss a few of the common filtering structures found in DSP applications implemented with reconfigurable computing.

The most common filter structure, the finite-impulse response (FIR) filter, was introduced earlier in Section 5.2.1. The FIR structure is an inner product of N filter coefficients with N samples of a signal. As illustrated in Figure 5.3, FIR filters can be implemented using a very parallel approach or they can be constructed using more iterative approaches that employ one or more MAC units. In the fully parallel approach, full multipliers can be replaced with smaller, more efficient constant-coefficient multipliers, assuming the filter coefficients are static for an application. Early references to FIR design for FPGAs can be found in [150, 290, 326]. A few more recent examples can be found in [64, 116, 126, 251, 257].

Unlike the FIR, the infinite-impulse response (IIR) structure employs feedback to implement a filter. The IIR filter may take one of many different forms. Figure 5.9 illustrates one example. Because of the use of feedback, IIR filters may require less hardware than FIR filters but they also require more effort to properly design. For example, the use of feedback makes it more challenging to understand the arithmetic precision required to avoid overflow conditions. As with the FIR filter, constant coefficient multipliers may be used to make the implementation more efficient. A few early and more recent examples of IIR structures on FPGAs can be found in [240, 276, 335, 339].

Though many applications use filters with static coefficients and thus can optimize the hardware appropriately, more advanced filtering structures,

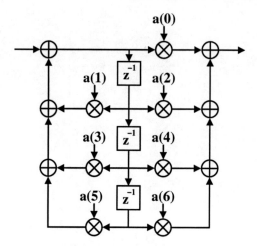

Fig. 5.9. IIR Filter Example

called "adaptive filters", are designed to automatically update their filter coefficients to improve their filtering capabilities as their inputs change over time. An example of this approach using FPGAs can be found in both [392] and [433], which provide comparisons of several implementation variations of adaptive filters that use the least-mean-square (LMS) algorithm to modify the coefficients of an FIR filter. Of the adaptive filtering applications reported in the literature over the last decade, most adaptive filtering approaches have used a variation of the LMS algorithm (e.g., [92,112,264,274]), though they are not the only approaches that have been used or proposed (e.g., [9, 164, 404]). This bias toward the LMS algorithm is likely due to the fact that the LMS algorithm requires less computational complexity than many other techniques and can be implemented without the use of floating-point arithmetic.

Finally, one popular structure used frequently for signal decimation (i.e., reducing the sample rate of the signal) is the cascaded integrator comb (CIC) filter [212]. Illustrated in Figure 5.10, this filter is popular because it can provide large sample rate changes, it does not require multiplication, and it requires relatively little data storage. When a (CIC) filter is used for decimation, truncation or rounding can be used effectively to control the growth of data precision, reducing hardware requirements. Notice that a CIC decimator (or downsample) filter is used in Xilinx's digital down converter illustrated in Figure 5.4.

5.3.3 Transforms

Much like the variety of different filters that are used, a number of different transforms are frequently used for DSP. Transforms convert a signal from one domain or representation to another. This is done when it is easier to manipulate the signal in the transformed domain. For instance, if the frequency con-

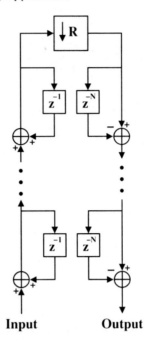

Fig. 5.10. CIC Filter Example

tent of a signal is desired, a Discrete Fourier Transform (DFT) is generally performed to convert a time-based signal representation into a frequency-based representation. We will mention only a few of the most common transforms, describing their basic operations and referencing related FPGA implementations.

Fast Fourier Transform

The DFT is probably the most common transform used in DSP. The equation for computing a DFT for a given input sequence is:

$$X(k) = \sum_{n=0}^{N-1} x[n]e^{-jk(2\pi/N)n} \tag{5.2}$$

The Fast Fourier Transform (FFT) [94] is an efficient and popular method for computing a DFT. As an example, when using a radix-2 FFT, only $Nlog_2N$ complex multiplications are required instead of the N^2 complex multiplications in the direct computation of the DFT—a significant savings. Though many radices are possible, we will use the radix-2 algorithm as simple example. In the radix-2 algorithm, the basic computation kernel, called a butterfly, is illustrated in Figure 5.11 in two of its forms. The basic form shown in (a) requires two complex multiplications and two complex additions using the complex coefficients $W_N^k = e^{-j(2\pi/N)k}$ and $W_N^{k+(N/2)} = e^{-j(2\pi/N)(k+N/2)}$; this, of course, translates into eight real-valued multiplies and eight real-valued additions. The modified form in (b) reduces the core computation to a single complex multiply, a complex addition, and a complex subtraction—equivalent to four real-valued multiplies and six real-valued additions. An early example of a radix-4 FFT implemented on an FPGA can be found in [326].

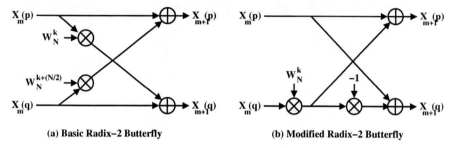

(a) Basic Radix–2 Butterfly   (b) Modified Radix–2 Butterfly

Fig. 5.11. Radix-2 FFT Butterflies

Noting that the complex multiplications in the butterflies are effectively just rotations in the complex plane, CORDIC units have frequently been employed to replace the complex multiplications—a significant savings considering that the unrolled CORDIC unit requires about the same resources as a single real-valued array multiplier. On the downside, the built-in CORDIC gain must be considered when taking this approach. Several FPGA implementations have taken advantage of this approach, including [35,121,353,421,438].

Discrete Cosine Transform

A transform frequently used in image processing (such as with MPEG video compression), the Discrete Cosine Transform (DCT) is another transform that has been regularly implemented using FPGAs and reconfigurable hardware. The one-dimensional DCT is defined by the following equation:

$$Y_k = \alpha_k \sum_{n=0}^{N-1} x_n \cos(\frac{2\pi}{4N}(2n+1)k) \tag{5.3}$$

where $\alpha_0 = 1/\sqrt{N}$ and $\alpha_k = \sqrt{2/N}$ for $1 \le k \le N-1$. The DCT has been popular for image compression, such as that used in earlier MPEG and JPEG standards, because of its excellent energy compaction properties for highly correlated data [229]. Because of the interest in image compression, most of the transforms reported in the literature are two-dimensional. Two dimensional DCTs, like two-dimensional FFTs, are often decomposed into computing two series of one-dimensional transforms—a series down the rows of the original image and another series down the columns of the row-transformed image.

A number of approaches for computing the two-dimensional DCT using reconfigurable hardware have been proposed and implemented. A few examples include [117], which compares a polynomial-based evaluation of the DCT to a straightforward distributed-arithmetic-based approach, and [155], which describes an approach for performing DCTs in such a way as to balance speed, cost, power, and precision-induced error. These techniques require little or no multiplication, reducing the size and improving their suitability for FPGAs. Examples of bit-serial, digit-serial, and distributed-arithmetic approaches to the DCT are described in [153, 238]. Unlike with the FFT, the DCT implementations above do not use a common kernel for their computations.

Discrete Wavelet Transform

The final example of a common transform used in reconfigurable computing for DSP is the Discrete Wavelet Transform (DWT) [278, 344]. DWTs have been used for a number of signal processing applications. Probably one of the more recent notable applications is for JPEG2000 image compression. Wavelet transforms offer some features that other transforms, such as the DFT, do not. For instance, the DWT can offer time-resolution information and can handle transient signals better than the DFT. Also, the DWT can represent the scale of signal features. Early implementations of the DWT on FPGAs are mentioned in [358, 398].

Like the DFT and DCT, the DWT transforms a time series of samples (or an image) into a sum of weighted basis functions representing the time series (or image). The DCT and DFT use a set of sinusoidal basis functions, but the DWT uses a more unique set of basis functions, called wavelets. Figure 5.12 illustrates an example of a four Haar wavelet basis functions (taken from [282]). Note that these basis functions are defined for a finite duration of time (hence the name wavelets) and that all of the wavelets are based on time shifts as well as dilations or contractions of the original "mother" wavelet. The finite duration of the wave makes the implementation of the DWT a simple FIR-style computation.

One of the advantages as well as complexities of wavelet transforms is that the set of actual wavelet basis functions that are used is actually a parameter, unlike with the DFT and DCT transforms. This can be an advantage because the basis functions of the DWT can be optimized to better fit some application or implementation method (e.g., using only integer arithmetic). To be able to produce hardware for any DWT or many different DWTs, some parameterized method for generating the hardware is desirable. An example of an approach to this problem for FPGAs is given in [310].

Several approaches for implementing DWTs in hardware, of course, have been developed. We will mention a few. The pyramidal algorithm described in [278] is a common algorithm for implementing wavelet transforms and takes advantage of the fact that the DWT can be defined recursively. Figure 5.13 provides examples of this filter bank approach for both the forward and inverse

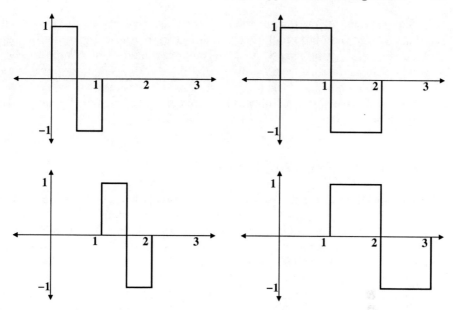

Fig. 5.12. Four Haar Wavelets

DWT . Note that pairs of quadrature mirror filters and either downsampling (for the forward transform) or upsampling (for the inverse transform) are used. The filters themselves are implemented as FIR structures. The FPGA implementations described in [7, 44, 323] are variations of this pyramidal algorithm.

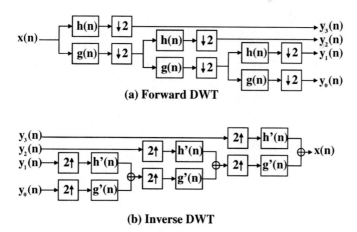

Fig. 5.13. Pyramidal Algorithm for Discrete Wavelet Transform and Its Inverse

The common algorithm mentioned above has several drawbacks. It can require a large amount of buffering for intermediate results, the processing latency can be large, and results in a high computational cost. Other approaches based on lifting-factorization (as illustrated in [24, 25, 282]) use a finite state machine representation of the DWT and allow the transform to be performed in place while allowing for a parallelized computation that requires relatively little communication between computational blocks.

5.4 Example DSP Applications

As mentioned in the earlier sections of this chapter, FPGAs and reconfigurable hardware are being applied to more and more DSP applications due to the speed-up provided by application-specific hardware solutions. In some cases, such as in low- to medium-volume applications, FPGAs are taking over the role once played by ASICs as fixed application specific hardware due to the increased cost and time-to-market ASICs require.

Beyond these fixed function applications, reconfigurable hardware can be applied in DSP applications where reconfigurability is beneficial or even required. We will briefly describe a DSP algorithm and an application that fit this description, namely, beamforming and software radio.

5.4.1 Beamforming

Beamforming [401] is a spatial (as opposed to temporal) filtering operation that combines information from an array of sensors to identify a signal's direction of arrival. It can amplify signals from specific directions relative to other directions through simply processing the data appropriately. Basically, beamforming uses the fact that identical signals that are in phase with each other add constructively to produce a signal with the maximum amplitude while those that are out of phase will provide a signal of much smaller amplitude due to destructive interference. Though rarely a complete application by itself, it has been a key component requiring flexibility and high performance in applications areas ranging from sonar and radar to wireless communications and software radio.

To perform this spatial filtering operation, the signals themselves can be summed in the time domain or the frequency domain (i.e., after an FFT has been performed). In the time domain approach (sometimes called "delay-sum" beamforming), a history of the last N samples are kept for each sensor and a beam is formed by selecting the proper sample from each sensor history buffer and then summing the samples across the sensors. The sample chosen from each sensor's history buffer is determined by the angle of arrival for the particular beam and, hence, the relative delay expected between sensors as the signal propagates over the sensor array. For more realistic delay-sum beamformers, a windowing function is applied to the selected samples to better shape the beam response—thus, the basic operation becomes a multiply-accumulate operation. Though the computational operations themselves are simple, the real challenge often is effectively using memory bandwidth to provide the needed samples to the MAC units for the directions of interest.

In [179], a sonar beamforming application having 10,000 beams, a 2-kHz sampling frequency, and 400 sensors is described. The effective processing rate required for real-time processing of the data is $16x10^9$ $operations/second$. In comparing FPGAs with a high-end commercial DSP of the time, the FPGAs provided six to twelve times the processing capability of the DSPs. The data parallelism that was exploitable, the pipelined operation of the RC hardware, and the available memory bandwidth in the reconfigurable computing solution were significant factors in this speed up.

In the frequency domain, an FFT is performed on the time domain samples for each sensor, then phases value are added to or subtracted from the FFT data from each sensor for the frequency bins of interest and for each of the directions of interest. Figure 5.14 illustrates a "brute force" frequency domain beamforming operation for sonar from [180] that has 10,000 beams, 256-sample history buffers for each sensor, and 400 sensors.

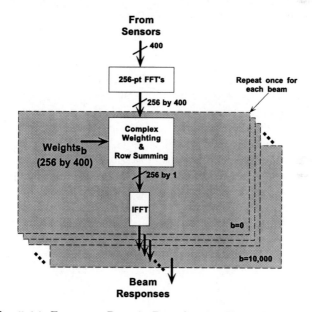

Fig. 5.14. Frequency-Domain Beamforming Example for Sonar

These phase values can be added in several ways. If the complex FFT data is in rectangular coordinate format, the rotation can be performed directly using a complex multiplication at the cost of four scalar multiplications and two scalar additions. If the FFT data is in a magnitude-phase (i.e., polar) representation, the phase can be simply added, but a polar-to-rectangular conversion must be done before summing across the sensors. Figure 5.15 provides a few examples of frequency-domain beamforming kernels described in [180]. Note the use of CORDIC to directly perform polar-to-rectangular conversions. Further, to reduce the storage requirements for the phase shifts for each frequency and beam direction, the kernel in Figure 5.15(c) has been modified to compute the phase shifts dynamically based on the time delay and frequency. As reported in [180] , the frequency-domain kernel computation required 8-16x more PDSPs than Xilinx FPGAs when comparing contemporary devices for providing real-time beamforming for the problem illustrated in Figure 5.14. Again, The data parallelism that was exploitable, the pipelining, and the available memory bandwidth in the reconfigurable computing solution were significant factors in this speed up.

A special application of the frequency domain beamforming technique—a matched-field beamformer—is described in [224]. This sonar beamforming algorithm accounts for the multiple paths a single signal can take through an ocean environment to provide three-dimensional location information for the signal source. In [224], the authors compare contemporary general-purpose computing solutions with a reconfigurable computing solution. The reconfigurable computing solution—a single SLAAC1 board using Xilinx XC4000 FPGAs—provided a computation rate of $3.8\,Gops/s$ for the problem, outperforming the general-purpose computing systems by a factor of 18 to 83 times. Using 2002 technology rather than SLAAC1's 1998 technology, a Virtex-II reconfigurable computing solution employing an Osiris board provided a computation rate of $60\,Gops/s$ for this same problem.

In addition to the above techniques, [231, 316] and other texts describe a number of adaptive beamforming algorithms that allow the spatial filtering operation to adapt to its environment to be more selective for the directions of interest (e.g., cancel jammers from other directions). The following are a sampling of the papers discussing adaptive beamforming employing FPGAs and reconfigurable computing: [226, 267, 279, 295, 315, 409, 410, 435].

Considering the large number of beamforming algorithms and techniques available, reconfigurable computing can provide a high-performance digital signal processing platform for beamforming while providing the flexibility to support multiple beamforming algorithms for a given sensor array. This flexibility, of course, enables the processing to be changed, as necessary, to meet specific application demands. For instance, the RC hardware can be put into a mode to search for signals using a straightforward beamforming algorithm. Once a signal of interest has been identified, the RC hardware might be configured to use an adaptive beamforming algorithm to help track a specific signal or better amplify the signal from a given direction. Further, using a reconfigurable computer for beamforming may allow the same processing hardware to support many different sensor arrays.

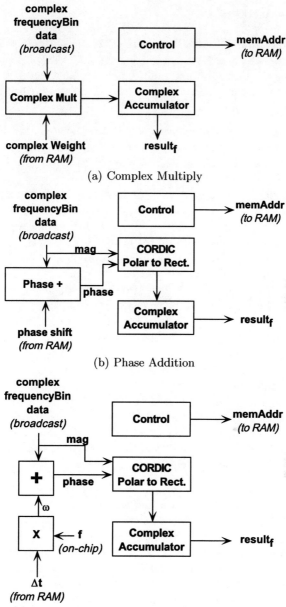

(a) Complex Multiply

(b) Phase Addition

(c) Phase Addition with Phase Shift Computation

Fig. 5.15. Frequency-Domain Beamforming Data Paths

5.4.2 Software Radio

Software radio [292] (SR) has been one of the most significant DSP applications for reconfigurable computing since the late 1990's. A software radio is a radio that can have its functionality substantially modified by software or hardware reconfiguration. The desire for software radio has been driven by many factors, including the following [341]:

- *Multifunctionality:* For example, a single device could be used for both Bluetooth and 802.11 wireless networking and support the optimized transmission and reception of different data types (voice, text, still images, video, networked game play, etc.).
- *Global Mobility:* A single device could be used across the globe despite the many various wireless communication standards. Likewise, large military organizations, such as the United States military, that have many different internal communication standards would be able to use a single device to seamless communicate with equipment using any of these standards.
- *Compactness and Power Efficiency:* Instead of needing separate hardware for each wireless communications standard, a single device can perform the function, potentially reducing the hardware and, thus, power requirements.
- *Ease of Manufacture:* As mentioned in the introduction to this chapter, reproducing analog systems with precision can be challenging compared with digital systems.
- *Ease of Upgrades:* The complexity of communication systems and their data handling can easily lead to flaws in the system that would be expensive to fix if found late in the design flow, for example, in the field by customers. Such flaws could be fixed in the field with a software radio architecture. Further, as new standards develop and improved algorithms are discovered, the customer's radio can be upgraded to support additional and/or improved functionality.

Figure 5.16 illustrates a general SR architecture. A software-defined radio receiver uses analog front-end hardware for providing the interface to the antenna, analog-to-digital conversion to convert the incoming signals to the digital domain, and then uses digital signal processing to extract the information from incoming signals. Conversely, if the unit can perform transmission, information is embedded in a signal digitally, the resulting digital signal is then converted to analog, and then back-end analog electronics are used to interface to the transmission antenna to transmit the signal. Note that in addition to DSP functions that perform channelization, sample rate conversion, and signal demodulation, software radios often include additional functions, such as error correction, data encryption, and data decryption.

Reconfigurable computing has been explored as an implementation technology for the digital signal processing portions of software radios due to its reprogrammability and high performance. Considering the numerous publications on software radio with reconfigurable computing, the following para-

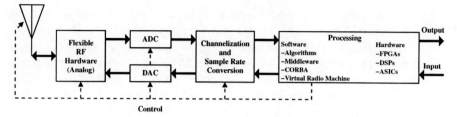

Fig. 5.16. General Software Radio Architecture based on [341]

graphs will provide only a brief overview of the reconfigurable computing literature discussing software radio.

Systems and Architectures

To begin our survey, we will first describe several papers and articles that deal with software radio systems and architectures involving reconfigurable computing.

One of the more complete software radio architectures supporting reconfigurable hardware has been described in [377, 378, 380, 381]. This architecture has four layers: the application layer, the soft radio interface layer, the configuration layer, and the processing layer. The application layer is the top-level application that provides the user interfaces with the radio. The soft radio interface (SRI) layer provides the application with the ability to select the function of the radio and to coordinate data I/O. The SRI layer packetizes incoming data and commands for the lower layers and stores all algorithms needed for the radio. The configuration layer manages the configuration of the processing layer and maps the algorithms from the SRI layer to their low-level configuration data, allowing the SRI layer to operate at a higher level of design abstraction. The configuration layer passes configuration and other control packets along with data packets to the processing layer. The processing layer reconfigures its operation based on command packets and operates on the incoming data packets. Some other notable features of this architecture include: data and control information are communicated using the same busses; the architecture supports pipelined, stream-based processing; and the communication of data among processing elements can be bi-directional when feedback is necessary.

In working with this architecture, researchers at Virginia Tech discovered that their Stallion reconfigurable computing system [380] based on a coarse-grained reconfigurable architecture supports this layered, streaming architecture quite well. For instance, Stallion specifically supports stream-based processing and provides wormhole reconfigurability where configuration data can be added as headers to data streams sent through the architecture. The layered architecture and Stallion have been used to build Code-Division Multiple Access (CDMA) receivers (e.g., [377, 378, 380]).

In [184], Gray et al. suggest that software radio architectures must be kept "object-oriented" to allow for easier reconfiguration both on the software and hardware sides. Further, their suggested architecture uses "interface" objects that are used to handle data-rate and data-format changes between hardware and software objects. By using their modular approach, they note that changing, for example, the modulation or filtering for a specific receiver channel can be done without affecting other channels, providing a very flexible architecture. Their goal is a system that can perform dynamic reconfiguration in order to handle changes in the environment (weather, radio motion, multipath) as well as being compatible with various networks. They briefly describe a software radio system they developed that supports binary phase shift keying (BPSK) and was implemented using a PowerPC microprocessor and Xilinx FPGA.

In [342], Revés et al. describe a DS-CDMA subsystem for software radio targeted to a reconfigurable platform. They provide the hardware costs for mobile-terminal and base-station functions and note the difficulty in developing software radios using FPGA design tools. In recognizing this difficulty and the difficulty of using heterogeneous processing resources in software radio systems, Revés et al. later developed a hardware abstraction layer for software radio applications so that an application could be mapped to general-purpose processors (GPPs), programmable DSPs (PDSPs), or FPGAs seamlessly [343]. The abstraction layer provides support for: seamless communication with and among processing resources, monitoring and control interfaces, real-time synchronization among processing resources, and methods for launching and mapping application objects to processing resources. In [343], they report on their implementation of this layer for FPGAs, noting that such a scheme is better suited for systems with larger FPGAs, where the area overheads would be lower. In their example system, the area overheads were 56% for the base station and 72% for the mobile terminal.

A number of software radio systems use reconfigurable hardware for the high-bandwidth front-end processing and other more traditional processors for lower bandwidth tasks. This is often due to the relative ease of developing software for GPPs and PDSPs and, sometimes, due to the need for floating-point in later processing stages. For instance, in [368], a hybrid system is presented that consists of a GPP, a PDSP, and an FPGA processor and provides a software radio solution supporting both the Japanese PHS wireless communications standard and the IEEE 802.11 wireless LAN standard. In this solution, the GPP provides the media access control layer and higher functions while the PDSP and the FPGA share the processing load for the physical layer for the wireless LAN.

A large number of papers and articles, of course, describe specific software radio implementations using reconfigurable hardware technologies. The following are just a few examples. In [242], Korah and McDonald describe a software radio design that uses an adaptive antenna array (i.e., adaptive beamforming) receiver to outperform the commonly used RAKE receiver for a wideband CDMA application. Due to FPGA resource limitations, all but the beamforming was implemented in an FPGA. Harumaya et al. in [200] describe a mixer-less architecture using a direct-conversion receiver, ADCs, FPGAs, a GPP, memories, and DACs. The paper describes how digital processing is used to compensate for the short comings of the direct-conversion receiver technology and also describe how they support five modulation formats within the system. A very brief description of a CDMA software radio receiver system is provided in [281]. Despite the brevity they do provide information regarding how they parameterized the design as well as some statistics regarding the receiver's power consumption (< 1 W for both dynamic and static power) and some area estimates ($< 88\%$ of a Xilinx XCV300E). As a final example, Hwang and Chu in [225] describe a very specific form of QPSK receiver they created using Altera Stratix. By utilizing a Stratix device and its DSP blocks, the design used relatively little logic (4%) and on-chip memory (2%) while using 65% of the on-chip DSP blocks.

A few papers have been written describing software-radio capabilities for space-based platforms. Caffrey et al. describe their reconfigurable computing system for space-based software radio in [61]. The system supports the processing of two separate input channels and provides support for mitigating FPGA programming data upsets due to radiation effects. In [319], the researchers describe the need for software radio in multimedia satellite applications and discuss a few ideas regarding the protocols and hardware support needed.

Finally, with regards to reconfigurable computing systems for software radio, [379] provides a nice evaluation of many reconfigurable hardware systems for their suitability to provide low-power software radio handsets. Some of the features that were identified as desirable included:

- support for hardware paging to better utilize the hardware;
- fast reconfiguration times ($< 0.2s$) to avoid "dropped" calls due to reconfiguration;
- static hardware for frequently used functions (multiplication, filtering, etc.);
- coarser granularity than FPGAs;
- scalable architectures that easily allow algorithms to be implemented across multiple devices or reconfigurable computers;
- strategically placed shift-registers, circular buffers, and other needed memory structures; and
- a large number of I/O pins for data throughput.

In their survey, Srikanteswara et al. also recognized several challenges for using reconfigurable computing hardware for software radio. These included:

- the need for better high-level design tools and compilers;
- the need for designs tools to represent and manage reconfiguration explicitly (e.g., the tools should minimize reconfiguration time in addition to execution time);

- very-low-power implementations will be needed for software-radio handsets[2];
- the need for easy integration with PDSPs, assuming that flexible, event-driven processing will be necessary;
- the need to carefully profile software radio designs to understand which parameterizable cores to use in coarse-grained architectures; and
- the need to manage the various run-time control parameters for the optimum implementation while allowing for changes to a system's algorithms and the environment in which the handset operates.

Reconfigurability

Of course, since one of the main benefits of reconfigurable computing is reconfigurability, a number of software-radio-related articles address the reconfiguration process directly. A number of the papers and articles mentioned above address reconfigurability. For instance, the layered software radio architecture described in [380] specifically addresses reconfiguration as part of the system, having a layer for configuration management. The three papers addressing space-based software radio and applications— [61], [184], and [319]—briefly address reconfiguration and the system needs to support it. As another example, the hardware abstraction layer of [343] explicitly accounts for reconfiguration in the API it provides.

In addition to the previously mentioned papers, Honda et al. in [213] consider an efficient method for transmitting configuration data for components of an FPGA-based software radio such that the radio performance is not significantly reduced. The researchers assume that a reduction in the radio's performance may result from errors in the the configuration data that the radio receives in addition to the usual radio transmission noise that radios experience. To achieve a 25% reduction in the number of symbols used to transmit the configuration data without significant affects on the radio, the researchers use two different modulation schemes for configuration data transmission: they use a more reliable but expensive scheme for the most significant bits of functions (adders, multipliers, etc.) while using a more efficient, but less reliable scheme for the least significant bits of the function. They assume that LUTs are used to perform logic on the FPGA.

Finally, in [110], the researchers consider a hybrid DSP-FPGA platform for software radio. In their system, they were able to illustrate the performance benefits of FPGA partial reconfiguration for software radio applications by reducing FPGA reconfiguration time by 45% through partial configuration.

---

[2] Reconfigurable hardware currently provides power efficient implementations for high-performance systems, but this is not the same as providing a very-low-power design.

DSP Functions

A software radio requires many signal processing functions to perform successfully. Many of the papers mentioned above describe some specific design details for implementing specific software radio functions.

As a few additional examples of what has been done, Chris Dick and his colleagues have published several papers on efficient FPGA implementations of software radio components. These include the parts of an OFDM receiver [118], an adaptive channel equalizer [120], and the hardware for carrier and timing synchronization [119, 122] for FPGA-based software radios.

Also, in [272], Lund et al. discuss a flexible, reconfigurable convolutional decoding system using FPGAs. Specifically, they describe how to implement a parameterizable decoder that can be reconfigured to decode any convolutional code up to constraint length 9 and at any rate to a minimum of 1/6. Though error control coding has not been emphasized as a digital signal processing function in this chapter, it does play a very significant role in most forms of communications.

Finally, Kim et al. in [237] describes an adaptive antenna signal processing algorithm for improved direction-of-arrival estimation (i.e., adaptive beamforming). In the article, they describe their algorithm, which performs eigenvalue decompositions using CORDIC, and they discuss the estimated size and performance of the circuit as well as the algorithm's performance when using an fixed-point implementation.

5.5 Summary

Digital signal processing has become a popular method for processing the signals we encounter in our daily lives and in other settings and reconfigurable computing has become a favorite technology for implementing DSP applications due to its flexibility and performance. The key to the performance of DSP applications on reconfigurable computers is the large amount of parallelism that can be exploited in these applications. Factors such as pipelining, small data widths, memory bandwidth, I/O bandwidth, and others help determine how much parallelism can be exploited and how efficiently the reconfigurable computing application operates.

We have also provided a small sampling of the wide variety of DSP operations and applications that have been developed using reconfigurable computing. Software radio using reconfigurable computing has been one active area of research and we expect the efficient use of reconfigurable computing for DSP to continue as an active research topic for some time.

Image Processing
Reid B. Porter

The goal of image processing is to robustly extract useful, high-level information from images and video. The type of high-level information that is useful depends on the application. Examples of applications include object detection and tracking for surveillance, defect detection for automated production systems, and scene classification for remote sensing and map annotation. Extracting high-level information is a difficult research problem and many different algorithms have been suggested.

6.1 RC for Image and Video Processing

Image and video processing have been significant application drivers for the reconfigurable computing community since its inception in the early 1990's. Prior to modern reconfigurable devices, image and video processing were also significant application drivers for computer architecture research and VLSI design. There are two related reasons why this application domain has received so much attention in the computer architecture community. The first is the poor, often unacceptable, performance observed in general-purpose processor implementation. This can be attributed to:

- large volumes of data
- exceptionally high memory bandwidth requirements
- real-time processing constraints

The second is the increased, often incredible, performance gains observed in custom or application-specific implementation. This can be attributed to:

- abundant parallelism in both data and algorithms
- local and regular data dependencies
- simple fixed point arithmetic and logic operations
- relatively small bit-widths

Reconfigurable computers have been used most widely, and successfully, for accelerating *low-level* image processing algorithms. These algorithms are typically applied close to the raw sensor data and are characterized by large data volume. Conceptually, low-level image processing is decomposed into a processing pipeline with raw image data (taken from a sensor) as input and the desired information as output. Figure 6.1 depicts a typical processing pipeline. Each stage of the pipeline can be a multiple-input, multiple-output transformation.

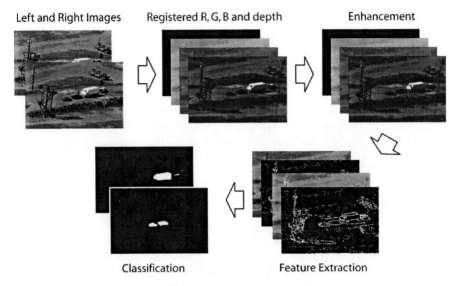

Fig. 6.1. A multi-spectral image processing pipeline

- There may be multiple sources of data from multiple sensors and / or multiple points in time in which case, it can be useful to co-register the data. The relative displacement between data sources is often useful, e.g., depth information can be recovered from stereo pairs, and motion information from temporal sequences.
- The image enhancement stage is concerned primarily with removing sensor noise and other environmental variation in order to make subsequent analysis easier.
- Feature extraction is a transformation from the image space, where each pixel usually represents intensity, to a feature space where pixels represent more abstract quantities. These quantities are typically application specific and are chosen to make subsequent processing easier.
- Latter stages of low-level image processing include detection, classification, and segmentation, in which abstract labels are assigned to image

pixels. These labels are typically application specific, e.g., a non-zero label specifies a region of interest.

Many variants and extensions of this processing pipeline exist for particular applications. In terms of reconfigurable computer implementation, it is useful to categorize low-level image processing algorithms based on their data dependencies. Two broad categories of algorithm are:

Local Algorithms: The algorithm depends on data from a relatively small (compared to the image size) neighborhood that is local in spatial and temporal dimensions. Examples include, point or pixel operators (such as band arithmetic, thresholding), convolution, and motion estimation.

Global Algorithms: The algorithm depends on data from the entire image. Examples include transforms such as the Fast Fourier Transform and principle component analysis as well as statistical histogram techniques.

A general rule of thumb for obtaining speed-up with custom computing architectures is to minimize the number of times the data is accessed. By definition global algorithms often require multiple passes through the data and performance compared to general purpose processors is varied and algorithm specific. In this chapter we will concentrate on local algorithms. These algorithms are found in all aspects of the low-level image processing pipeline and they can benefit greatly from RC implementation. Reported speed-ups are typically two orders of magnitude compared to general purpose processors.

6.2 Local Neighborhood Functions

Local neighborhood functions (also called sliding window functions and spatial filters) are used extensively in image processing and computer vision. These functions are applied at a particular pixel location and their output depends on a finite spatial neighborhood. The function is applied independently at all pixel locations and is typically constant across all pixel locations. Figure 6.2 illustrates the how a neighborhood function is applied for a 3 by 3 neighborhood. When local neighborhood functions are applied at edge locations some of the neighborhood is not defined. The undefined pixels can be assigned a value of 0, or can be assigned the value of the closest pixel. Another common approach is to temporarily increase the size of the input image by reflecting pixel values across each edge.

The neighborhood function of Figure 6.2 can be generalized in several ways to include a large number of standard image processing algorithms. Neighborhoods can be generated from multiple input images such as color channels or more generally, spectral dimensions. The kernel (see Section 6.3) is 3 dimensional and the neighborhood function slides over the 2 spatial dimensions of the image stack. Point operators such as band arithmetic (average of color channels), clipping, thresholding and pixel scaling can be considered local neighborhood functions when we assume a spatial neighborhood size of 1

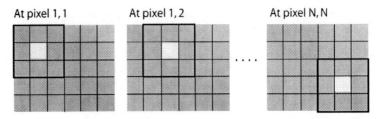

Fig. 6.2. A neighborhood function is applied to all pixels in parallel

pixel. These basic operations are described in detail in most image processing texts [229].

Local neighborhood functions can also receive multiple images over time, and this is typical in video processing applications. This is different from receiving multiple spectral inputs associated with a single image. Similar to FIR (Finite Impulse Response) and IIR (Infinite Impulse Response) filters encountered in signal processing the neighborhood window has a finite temporal extent and slides through time as the function is applied at each time step. The kernel is 3-dimensional and the neighborhood function slides over 3 dimensions (2 spatial and 1 temporal). Note that in the spatial dimension the neighborhood function is applied independently at every location, but for the temporal dimension this is not always the case as in neighborhood functions with temporal feedback (IIR).

Local neighborhood functions demand exceptionally high bandwidth to image data. For example, for a modest 3 band 256 pixel wide by 256 pixel high color video sequence, a (typical) 7 by 7 spatial neighborhood size and a 3 frame temporal window, the most general neighborhood function would require access to 441 pixel values at each image location. To obtain real time processing rates at 30 frames per second would require access to approximately 870 million pixels per second. As we will see, most image processing applications are composed of large numbers of local neighborhood functions and therefore the bandwidth requirement quickly exceeds what general purpose computing can provide. Fortunately, due to the regular nature of the memory access across the image array, there are also many opportunities to optimize the memory access. Reconfigurable computers are ideal platforms to tailor memory hierarchies and implement algorithm specific address generation, and therefore great performance gains are possible.

There are two main ways of achieving speedup in local neighborhood functions using reconfigurable computers: pixel parallelism (Single Instruction Multiple Data) and instruction-level parallelism through pipelining. The two extremes of this approach are illustrated in Figure 6.3.

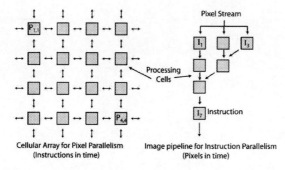

Fig. 6.3. Pixel Parallel versus Instruction Parallel

6.2.1 Cellular Arrays for Pixel Parallelism

Cellular arrays naturally model image data [349]. They consist of an array of cells in two, three or more dimensions. Each cell is associated with an image pixel and each cell has dedicated connections to its local neighborhood. This high-bandwidth local communication is ideal for implementing neighborhood functions; all pixels are processed in parallel, and the entire image is updated in 1 instruction cycle. FPGAs can implement a programmable, maximally parallel implementation of a cellular array, but can only efficiently implement a small numbers of cells. Large arrays require multiple FPGAs and/or time multiplexing, and the I/O required to initialize the array and read results can dominate the computation time.

6.2.2 Image Pipelines for Instruction-Level Parallelism

In this case only one cell of the equivalent cellular architecture is implemented. Data is provided to the cell through a continuous stream of pixels supplied to the cell one sample at a time, and usually in raster scan order. This arrangement is often suitable for real-time systems where data arrives directly from a serial I/O sensor. Since pixels are processed sequentially, the main way to achieve speed-up for an image pipeline is to execute multiple instructions in parallel. As shown in Figure 6.3 instructions can be implemented either in parallel (increasing the pipeline width) or in series (increasing the pipeline depth). Unlike cellular architectures, accessing a local neighborhood within an image pipeline must be carefully considered. All instructions in a pipeline are being executed at the same time, and therefore it may be difficult to provide data to all instructions at the right time.

We have described the two extremes of a pixel parallel verse instruction parallel design space. In practice any combination of these two extremes may be used. The optimal design point is dictated largely by the memory architecture of the particular FPGA and reconfigurable computer. Formalizing design choices and providing tools that can optimize among design choices is a topic of ongoing research [262].

6.3 Convolution

Perhaps the most well known local neighborhood function is convolution, which is defined in Equation 6.1. A multiplicative weight W is associated with each location of the neighborhood $\{k, l\}$, collectively known as the kernel K. The output F of the function is the accumulated weighted sum of the kernel applied at each pixel location $Image(i, j)$.

$$F(i, j) = \sum_{k, l \in K} W_{k,l} * Image(i - k, j - l) \qquad (6.1)$$

By selecting the appropriate weights, convolution can implement low-pass, high-pass and band-pass frequency domain filters used extensively in image enhancement and feature extraction. Low-pass filters use positive weights and are used for image smoothing. High pass filters use a kernel with a positive center weight and negative outer weights and are used to enhance high frequency components in an image such as edges and fine detail.

One of the first reconfigurable computer implementations of convolution was on Splash 2 [338]. The image pipeline approach was used in a linear systolic array implementation. Local memory was used to replace multipliers and the lack of on-chip memory meant the image width was limited to 32 pixels. Despite these limited resources the timing for a two 3×3 convolutions applied to a 512 by 512 image was 100 frames per second. A 3×3 convolution implementation which is very similar to the linear systolic array is shown in Figure 6.4. The image data is assumed to arrive one pixel each clock cycle in raster scan order. After a fixed latency, this architecture provides access to the entire neighborhood of data every clock cycle.

The architecture in Figure 6.4 places the lowest demand on external memory bandwidth, but the highest demand on internal memory bandwidth. Each pixel in external memory is accessed only once but for an image width, W, and kernel width, M, $((M - 1) \times W + M)$ pixels must be stored on-chip. Since around 1998, many modern FPGA devices have a the large amount of on-chip memory and this approach has been widely adopted [230], [303], [45].

The length of the shift register in Figure 6.4 depends on the input image. If the image is thousands of pixels wide it is unwise to buffer the entire row. The most common approach is to choose a row length appropriate to the hardware resources at hand (e.g., 64, 128 or 256 pixels), slice the input image into strips of this width, and provide these strips as one long, narrow image to the hardware. Due to the neighborhood, these strips must overlap by a particular number of pixels in order to produce results that *stitch* correctly. This overlap leads to a slight decrease in performance compared to the full

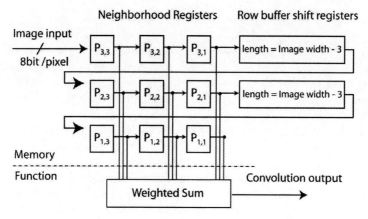

Fig. 6.4. A systolic array for 2-D convolution

length row buffers. Bosi, Bosi and Savaria estimate that dividing a 1024 by 1024 image into 16 slices reduces the number of registers required by a factor of 14.8 for a 3 × 3 convolution, while performance is reduced by 6% [56].

When on-chip memory is not available, row-length shift registers may not be possible at all. To maintain the pipeline throughput at one convolution per cycle the design needs to access more than one pixel per clock cycle from external memory. If we can access M neighborhood pixels per cycle we can do without the row length shift registers entirely. For example, if the data-width of external memory is 32 bits and the pixel data width is 8 bits, we can access up to 4 pixels per clock cycle. For a 3 by 3 convolution we need to access three pixels from three different scan lines each clock cycle. Since an image is typically stored in raster scan order in the external memory the memory access must cycle between the three different scan-lines. On-chip registers can be used to buffer the three consecutive pixels from each row and maintain throughput at one convolution per cycle [56].

In the situation just described each pixel is read from memory three times (once for each row in the neighborhood). To reduce the redundant I/O it is possible to implement multiple neighborhood functions, each associated with consecutive rows of the image. The multiple functions exploit pixel parallelism, and also, share local neighborhood access and therefore the I/O is reduced. This approach is described as partial loop unrolling by Draper et al. with respect to the Single Assignment C compiler [127].

6.4 Morphology

The pipelined neighborhood cache in Figure 6.4 can be used for a much wider class of algorithm than just convolution. Mathematical Morphology defines a large family of image processing algorithms, which essentially replace the weighted sum function block with a neighborhood order statistic. The kernel for morphological spatial filters is also called a structuring element or region of support and defines the set of pixels from which an order statistic is derived. The shape of the structuring element is very important. The simplest filters are erosion and dilation. Erosion is defined as the minimum from the set of pixels defined by the structuring element and dilation is the maximum. Another popular morphological filter is the median.

Morphological functions are generally far cheaper to implement with digital logic than convolution type functions. First, morphology avoids the multipliers that can become expensive for large neighborhood convolutions. Second, order statistics such as maximum, minimum and median are closely related to the digital domain. The relationship is described by a technique known as threshold decomposition, which was first introduced to analyze the median filter [154]. Threshold decomposition allows gray-valued pixel images to be processed with bit-level hardware and Figure 6.5 illustrates the technique for a 1 dimensional median filter. Pixel inputs are first thresholded at all possible quantization levels; producing a binary *stack* for each input whose height is equivalent to the pixel value. Each quantization level is then processed independently with a positive boolean function. Positive Boolean Functions (PBFs) are a subset of Boolean logic functions in which no input may be negated. To regain a gray-valued output we simply sum the binary outputs from each level. There is a one-to-one correspondence between a PBF and an order statistic, where each logical AND is replaced by a minimum and each logical OR is replaced by a maximum.

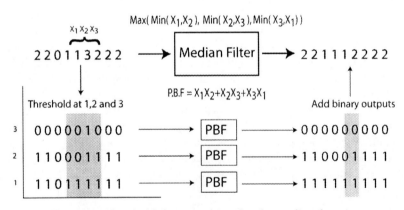

Fig. 6.5. Threshold decomposition for the median function

Threshold decomposition at first appears to have complexity proportional to the number of quantization levels (which may be very high) but this in fact can be reduced to the number of inputs in the filter window. Also, each stack (associated with both inputs and output) can only make a single transition from one to zero i.e. within the same stack ones cannot appear above a zero. This property, known as the stacking property, allows for extremely efficient implementations. Chen proposes a most significant bit first, bit-serial implementation which uses a single PBF [70]. When implementing a 3 by 3 8-bit pixel, erosion on a Xilinx 6200 series FPGA, Woolfries found that implementing 8 copies of the Chen's implementation used 75% fewer resources and was 33% faster than implementing the threshold decomposition in Figure 6.5 directly.

The threshold decomposition approach is particular useful in reconfigurable computing for implementing order statistics with high complexity, such as the median, and a large number of inputs. For simple order statistics, such as maximum and minimum the number of comparisons is linear in the number of inputs, and a direct sorting network can be implemented efficiently. The direct sort can also be used for the median function if the number of inputs is small [372].

6.5 Feature Extraction

We have described the basic neighborhood function building blocks used in image processing. By combining these building blocks in various ways we can implement a large number of more complex image processing algorithms that perform feature extraction. Feature extraction often has one of two aims:

1. To produce a representation that is invariant to specific image properties such as rotation, illumination, scale etc.
2. To produce a representation suitable for subsequent processing. These quantities often represent things like texture or color, but they can vary greatly depending on the application.

One of the most well known example of feature extraction is edge detection. Asymmetric weight kernels suggested by Roberts, Sobel, Prewitt and Laws estimate image gradients in specific directions. A number of these kernels are used in convolution and the outputs are combined to produce a rotationally invariant edge detection. Outputs are typically combined by the sum of squares, however a sum of absolute values or maximum may be more appropriate in RC implementations. Obtaining rotation invariance through multiple kernels is also used in morphology. Figure 6.6a shows an example where a linear structuring is used to *probe* the image for linear image features such as roads. A maximum is used to combine multiple outputs during dilation, and a minimum is used during erosion. For the image processing pipeline architecture, multiple rotations correspond to increased pipeline width. Considerable memory resources can be saved if multiple rotations share row buffers and neighborhood registers.

Gabor filters also need increased image pipeline width. The Gabor kernel is defined as a complex plane wave modulated by a Gaussian distribution.

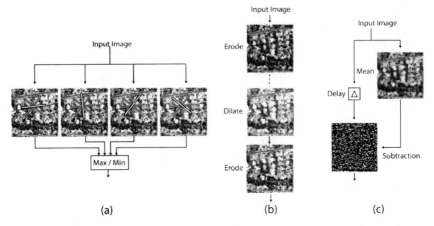

Fig. 6.6. Examples of feature extraction: a) rotationally invariant linear features b) deep morphological pipelines and c) adaptive thresholding

It implements a band-pass frequency domain filter. For feature extraction, a bank of Gabor filters are implemented, each tuned to specific spatial frequencies and orientations. The quantity used in subsequent processing is often the magnitude of the complex convolution which exhibits invariance to small shifts of the input image. The number of filters in a Gabor filter bank can be quite large, in which case, it may be more efficient to implement convolutions in the frequency domain. This requires a Fast Fourier Transform (FFT) and an Inverse FFT, which are available for most modern FPGAs as third party IP cores [406].

Many complex morphological algorithms for feature extraction such as opening, closing, open-close and close-open filters are built by successive application of erosion and dilation. Usually the shape of the structuring element is constant between successive erosions and dilations. As shown in Figure 6.6b these algorithms can be implemented in the image pipeline architecture by simply increasing the pipeline depth.

Another class of feature extraction algorithms are locally adaptive, which means that a neighborhood function is dependent upon some statistic of the local neighborhood. A popular example is adaptive thresholding in which the center pixel is thresholded by the mean or median value of the neighborhood. Locally adaptive functions define small sub-trees within an image processing pipeline. For implementation within the image pipeline, different pipeline paths must be latency adjusted before they can be combined pixel-wise. The adaptive threshold example is illustrated in Figure 6.6c.

There are many other examples in feature extraction where algorithms are implemented by cascading multiple local neighborhood functions. Reconfigurable computers gain a significant advantage over general purpose processors for these types of algorithms. Apart from increased latency (which in many applications is not important), the pipeline throughput is constant at one pixel per cycle. The FPGA resources limit how far this approach can be taken. Depending on the FPGA architecture and specific type of algorithm, this can be logic limited or memory limited. Once the limit is reached, multiple passes of the data or additional FPGAs are required to execute further instructions.

6.6 Automatic Target Recognition

One application where FPGA resources are often not sufficient to implement the maximal throughput pipeline is template matching for Automatic Target Recognition (ATR). In template matching the neighborhood function kernel is a small sub-image from the original image or from a related image and the neighborhood function calculates a distance metric, such as correlation, between the sub-image and the original image at all pixel locations. Typically there are a large number of templates, or kernels, and the problem is to find the template with the best match at each pixel location. In practical systems the number of templates often far exceeds what can be matched in parallel with reconfigurable hardware. For example, an ATR algorithm developed for synthetic aperture radar by Sandia National Laboratory has approximately 5700 templates associated with each target. With tens or hundreds of targets, it becomes clear that a practical implementation will require a number of passes.

The most efficient hardware utilization is gained by customizing the FPGA for each pass with the configuration bit-stream. This is often appropriate when each pass of the data performs a significantly different type of processing, however the approach can also be used to generate optimal specializations of a generic pipeline. For example, Chia et al. have produced an ATR system called Mojave that produces specialized matching circuits for different templates [75]. They call their approach partial evaluation, and it exploits several properties of the Sandia application:

- The templates are sparse so not all neighborhood pixels are involved in the correlation. Chia et al. estimate that for approximately half the templates this approach uses 5.8% of the resources used in a general purpose circuit.
- Many templates share common pixels and therefore share partial results in the correlation.

The Mojave system provides a number of CAD tools that can automatically perform the above optimizations for a given set of templates. The system matches 8 by 8 templates against a 128 by 128 video image and was able to achieve an improvement factor between 2 and 10 over the existing ASIC implementation. Device reconfiguration is an attractive approach to multiple pass image processing. The approach is unique to reconfigurable computing and it can lead to significant performance improvements. One disadvantage of the approach is that it depends on being able to rapidly reconfigure the

FPGA. The Mojave system is based on the Xilinx 4013PG233-4 FPGA which requires 30ms to reconfigure. In comparison, the FPGA processes 4 templates in parallel and takes 16ms for 1 pass. The net result is a system that takes 46ms to evaluate 4 templates.

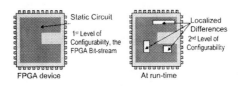

Fig. 6.7. Localized reconfiguration.

Another way to implement multiple pass hardware specializations is with partial reconfiguration. In many image processing applications, the image pipeline can have very similar implementation requirements from one pass to the next. Computations are typically regular which means implementation difference can be localized and reconfiguration time minimized. Figure 6.7 illustrates the concept. This approach was used for the Sandia ATR application by Bellows and Hutchings [43]. They targeted the rapidly reconfigurable Xilinx XC6200 series FPGA. Using placement constraints they arranged a 2-D systolic array of processors with static interconnect. At run-time the function of these processors is specialized based on a particular template that is being matched. The hardware efficiency of this approach can be very close to that achieved by a complete re-synthesis. The disadvantage of the approach is tied to the limited partial reconfiguration capabilities of most commercial FPGAs. For example, to manipulate the routing to select which inputs are supplied to a neighborhood function is difficult with most FPGA devices.

The alternative to FPGA based partial reconfiguration is to build the multi-pass variability into the hardware design itself. This involves increasing the complexity of the design to include the required variability, and provide on-chip configuration registers with an appropriate interface. Rencher and Hutchings used this approach when implementing the Sandia ATR application on the Splash 2. They implemented a single general purpose matching algorithm for a 16 by 16 template. Each template was loaded from local memory to on-chip registers where it was matched with the input image using a deep image processing pipeline. A control circuit monitored the match from each pass and maintained a record of the best template. Rencher and Hutchings estimated their design running at 13.2 MHz outperformed a HP 770 running at 110 MHz by two orders of magnitude. Building custom configuration circuits can be extended to any level of flexibility, e.g., configuration registers can store program instructions for an arithmetic logic unit or a micro-controller.

Typically, this approach will be most resource efficient when the multi-pass variability is localized and configuration circuits are tailored to the problem at hand.

6.7 Image Matching

Area based image matching is another class of local neighborhood algorithm that is used extensively in low level image processing. With stereo cameras, two cameras are used to image a scene from two different locations so that a physical point appears in different locations in each camera image. From the difference in location (called the disparity), the depth of the point can be calculated. In video cameras, object or camera motion produces a similar effect and the difference in location (called the displacement) can be used to estimate a motion vector. The image matching problem is to find the corresponding points in each image. In area based matching techniques, a point to be matched becomes the center of a neighborhood. The matching problem involves finding a similarly sized neighborhood in the second image that is the best match for the neighborhood in the first image. Figure 6.8a illustrates the matching problem for a single pixel. The procedure is repeated for every pixel in the template image.

Some popular metrics for matching include the sum of squared (SSD) and sum of absolute differences (SAD), as well as the normalized cross correlation (NCC):

$$F(i,j) = \frac{\sum_{(k,l) \in K} W_{k,l} * Image(i-k, j-l)}{\sqrt{\sum_{(k,l) \in K} W_{k,l}^2 * \sum_{(k,l) \in K} Image^2(i-k, j-l)}} \qquad (6.2)$$

Several metrics have been suggested that aim to provide the accuracy of NCC with less expense. A method that is particularly appropriate in FPGA implementations is to use the relative ordering of the pixel intensities to calculate similarity [436]. Images are first transformed according to local neighborhoods. In the *rank transform* each pixel intensity is replaced by an integer that represents the number of pixels within a neighborhood whose value is less than the center pixel. The *census* transform replaces each pixel with a bit string which encodes the neighborhood pixels according to their location. If a pixel value is less than the center pixel the corresponding position in the bit string is set to 1, otherwise it is set to 0. Once the images have been transformed, points are matched by using the traditional area based methods. The rank transform typically uses the SAD or SSD similarity metric while the census uses a metric based on the Hamming distance between the two bit vectors. We estimate the rank matching metric consumes approximately 50% fewer resources than SSD and at least 75% fewer resources than NCC. This is mainly due to the smaller data width of the rank metric output [329].

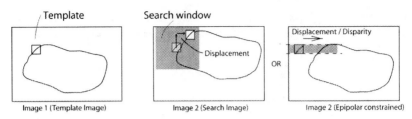

Fig. 6.8. Searching for the best match in a) the general case and b) in the epipolar constrained case

Image matching and template matching are in some ways similar. In both algorithms there are a very large number of templates, and the problem is to find the template with the best match. There are also two significant differences:

1. For image matching the template has a search window that is typically much smaller than the original image. In template matching each template is matched at every location in the entire image.
2. In image matching the templates are local neighborhoods taken from every pixel location in the template image, which means consecutive templates have overlapping values. In template matching each template may be completely different from every other template. When templates do have overlap (as in the Sandia application) it is template specific.

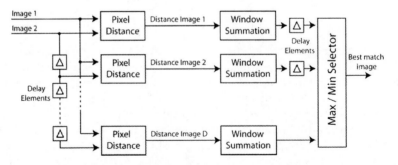

Fig. 6.9. A real-time matching architecture.

These differences suggest an alternative approach to implementation for image matching which is illustrated in Figure 6.9. The two image pixels streams arrive at the same time but then are displaced from one another by varying degrees via delay elements. Images can be displaced horizontally via registers, but require row-length shift registers to be displaced vertically. Often in stereo matching the two cameras are mounted carefully to ensure that the two scan lines are in correspondence. If this is the case then the displacement can be assumed to lie on the same horizontal line which greatly reduces the search window as illustrated in Figure 6.8b.

Most metrics used in matching, such as SAD, SSD, NCC and the Hamming distance, are based on a neighborhood summation which can be calculated in two steps. We first we calculate a distance image based on a pixel wise distance metric between the two displaced images. We then accumulate the distance image within a local neighborhood. This can be computed with the convolution function as in Figure 6.4, or since there are redundant additions (due to equal weights in the convolution), it can be computed with running totals [152]. Neighborhood summation with running totals allows much larger neighborhoods to be accumulated and is a two-stage process:

1. Calculate row sums: A new row sum is calculated from the previous row sum by adding the new pixel and subtracting the last pixel. The row sum calculation is easily pipelined with a neighborhood row shift register and an adder / subtracter.

2. Calculate column sums: This is similar to the first step but instead of accumulating and subtracting pixels we accumulate and subtract row sums. The number of running totals is equal to the image width. Since the row sums are being calculated in scan line order a large shift register is required to subtract the last row sum within the pipeline. One way to avoid this shift register is to introduce redundant additions so that the last row sum is calculated at the same time as the first row sum [130]. Figure 6.10 illustrates the main components required in calculating the column sums. The running totals are kept in memory and accessed sequentially as new row sums are generated.

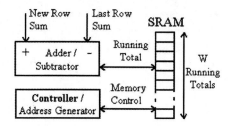

Fig. 6.10. Calculating the running totals.

To produce the final output, the neighborhood sums from the various distance images are compared. The displacement with the smallest sum (except for the NCC metric for which we choose the largest) corresponds to the best match in the search area. The implementation described has been used by several researchers to obtain real-time depth maps from stereo cameras [130], [420]. As far as we know the architecture has not been used for motion estimation on FPGAs. This is probably because of the large amount of memory that is required to search for vertically displaced neighborhoods, which until recently would have made real-time implementation infeasible. In addition, most RC implementations for motion estimation target block matching algorithms used in video compression [347]. In these algorithms, matching neighborhoods are usually non-overlapping which means only a subset of the pixels within the template image are matched. This leads to different opportunities for optimization and therefore different implementations.

6.8 Evolutionary Image Processing

In image processing we often define an error, or loss function that measures how well a particular algorithm solves the problem of interest. The task is then to find, through optimization, the algorithm that minimizes this loss function and is therefore in some sense *optimal*. Optimal image processing algorithms are generally able to outperform fixed algorithms since they are tuned for the specific data and task at hand. Many of the standard image processing algorithms, such as convolution, morphology and matching are used in optimal image processing, e.g., in optimal image enhancement we replace a fixed convolution kernel like Gaussian smoothing with convolution weights that are optimized to minimize a mean squared error. Another much studied optimal image processing problem is pattern recognition where the error function is based on detection and false alarm rates.

Reconfigurable computers are particularly useful in implementing optimization problems since the implementation requirements of optimization problems vary greatly from one problem to the next [4]. Evolutionary Algorithms (EA), define a family of optimization techniques for which this is particularly true. EA include genetic algorithms, genetic programming, evolutionary programming and evolutionary strategies. EA optimization is one of the most flexible optimization techniques in use today, and has been applied to a variety of research, industrial and commercial problem solving activities [105]. EA optimization is based on sample and test. A large number of candidate solutions are generated randomly. Each candidate is evaluated and assigned *fitness* by applying the solution to a training set and calculating the loss function. Based on this fitness, the population of candidate solutions is resampled, and the process repeats until candidates achieve a desired level of performance. EA can be applied to many problems in image processing but it is very computationally intensive. Each candidate evaluation is typically a complete pass of an image processing pipeline, and a large number of evaluations are required.

One of the most effective ways to use a reconfigurable computer for evolutionary image processing is as a fitness evaluator. The basic architecture is shown in Figure 6.11. An application specific image pipeline is implemented in much the same way as in conventional image processing. We then add a simple

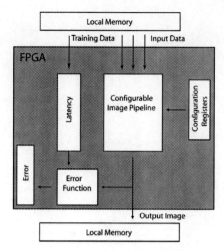

Fig. 6.11. A fitness evaluator architecture for evolutionary image processing.

control structure that compares the pipeline output to a desired output and calculates the error function. Note, if the pipeline exploits pixel parallelism it is likely that multiple error functions will be implemented in parallel. The pipeline must be used to evaluate many different candidates and therefore it requires a second level of configurability appropriate to the optimization problem. Similar to the ATR example, this can be implemented with partial reconfiguration, and/or with custom configuration circuits as illustrated in Figure 6.11.

Apart from fitness evaluation, the evolutionary algorithm itself is a very simple algorithm and can be implemented on the reconfigurable computing system. Sidhu et al. describe a genetic programming pipeline implemented on a XC6264 FPGA which obtains speed-up of 19 compared to a 200-MHz Pentium Pro for an arithmetic regression problem, and three orders of magnitude for a logic-based multiplexer problem [370]. For the image processing application domain the computation time for the fitness evaluation usually far exceeds the computation time of the EA. When the RC communicates with the host computer over an I/O bus, it is typically not necessary to implement the EA in hardware.

Using the reconfigurable computer as an I/O bus accelerator generally implies a low bandwidth connection between the reconfigurable computer and the host processor. To minimize communication across this connection it is important to calculate the error function on chip. Figure 6.12 illustrates the ideal arrangement. Large volume input data and training data are loaded once at the start of optimization to the RC local memory. Communication between host and the RC during optimization involves writing to on-chip pipeline configuration registers, initiating the pipeline evaluation, and then

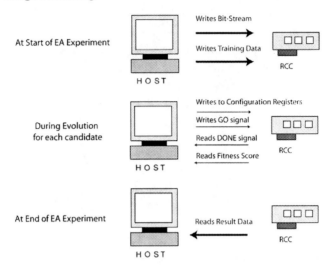

Fig. 6.12. Hardware / software portioning for evolutionary image processing.

retrieving the output error. Only at the end of optimization, is the result image from the lowest error pipeline retrieved for inspection.

Recently, in a field known as evolvable hardware, we observe an interesting consequence of approaching image processing as an optimization problem. The idea is to use evolutionary algorithms to explore non-tradition parameterizations of image processing problems to produce solutions with efficient FPGA implementations. One of the first applications of evolvable hardware to image processing involved optimizing a variable-length encoded PLD AND-OR array to solve a binary character recognition problem [228]. Many other examples of this approach have now been published [129], [432], [362]. Evolvable hardware researchers have developed many novel image processing algorithms by optimizing collections of low-level building blocks similar to FPGA logic cells. This approach can produce extremely compact solutions, but will produce little speed-up over general purpose machines unless a large number of these functions are implemented in parallel. This observation led us to develop a system called Pooka, which combines evolutionary image processing with a reconfigurable computer to solve scene classification and terrain mapping problems in satellite imagery [330]. In the Pooka system we explore a much more abstract parameterization of neighborhood functions, and focus on finding solutions that combine multiple copies of these functions within a deep processing pipeline. The net result is a system that can solve complex practical problems and obtain significant speed-up compared to a general purpose processor.

The Pooka pipeline is illustrated in Figure 6.13. There are 18 highly pipelined functions (or layers): 9 of these functions are used to combine multi-

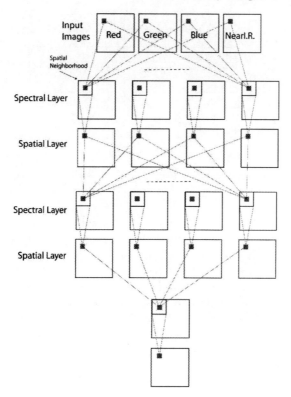

Fig. 6.13. A 18 layer image pipeline for multi-spectral image classification.

ple spectral channels and their spatial neighborhood is one pixel. The remaining 9 functions implement functions of a 5×5 neighborhood. The pipeline can have up to 16 different inputs. In the example in Figure 6.13 the input imagery has 4 spectral channels, but in multi-spectral imagery there can be many more. The connectivity at the pipeline input and between processing layers is made configurable through large multiplexers which are controlled by the on-chip configuration registers. The basic building block within the Pooka system has two inputs (a and b) and one output which are all 8 bit 2's complement integers. A three bit configuration register dictates which one of eight functions the building block implements. These functions are summarized in Table 6.1.

In each spatial layer there are 24 configurable building blocks. In each spectral layer there are 3 configurable building blocks. The connectivity between blocks is largely hard coded and is described with other implementation details in Porter et al. [330]. Pooka has been implemented on a Firebird reconfigurable computer from Annapolis Microsystems [227]. This is a 64-bit, 66-MHz PCI co-processor that contains a Xilinx Virtex 2000E FPGA, and a

Function	Operation
Average	$\frac{a+b}{2}$
Difference	$\frac{a-b}{2}$
Absolute Average	$\left|\frac{a+b}{2}\right|$
Absolute Difference	$\left|\frac{a-b}{2}\right|$
Maximum	$Max\{a,b\}$
Minimum	$Min\{a,b\}$
Select Left	a
Select Right	b

Table 6.1. The Pooka configurable building block.

total of 36 MB of on-board memory distributed in 5 independent banks. The 18-layer network used 64% of the FPGA logic and 35% of the block ram post place and route and is clocked at 50 MHz.

During evolution the Pooka system obtains speed-up of three orders of magnitude over a software simulation running on a 500-MHz Pentium III workstation. This is attributed to the fact that Pooka is based on a configurable building block that is efficiently implemented in hardware but inefficiently in software. In hardware, the configurable building block network is completely pipelined and is carefully hand designed to make the best use of Virtex FPGA resources. In software, the 24 configurable building block network requires many conditional assignments, all within nested loops. The software compiler has few optimizations available to it and the performance is poor. A possibly more meaningful measure of performance can be estimated by considering a high-level approximation of Pooka components. For each Spectral layer in the pipeline, a linear combination is calculated. For each Spatial layer, a 5 by 5 neighborhood average is calculated. The relative speed-up in this case was estimated at two orders of magnitude. However the software implementation is slightly simpler (but has greater bit-widths) than the Pooka implementation.

The speed-up achieved by RC in evolutionary image processing can be close to that achieved in real-time reconfigurable computing systems. This is because there is very little communication between the host and reconfigurable computer during optimization. Image data is always on-time (since it is stored in local memory) and the image pipeline operates almost continuously at peak capacity. However, once the optimization is complete, the performance of the optimized image pipeline in application depends on several factors which are typically related to data I/O. In the Pooka system, optimized pipelines are typically applied to large satellite images (2 to 4 Gigabyte images) that are stored on the host computer hard disk. Therefore, the execution time for the Pooka system must include the time required to read data and write results to disk, as well as the time to transfer data to and from the FPGA board across the PCI bus. We found the communications overhead reduced the 100X speed-up by a factor of 10. This overhead is not a factor when image frames are acquired and processed in real time.

6.9 Summary

The local, regular nature of local neighborhood functions, which are used extensively in image processing, provide many opportunities to exploit parallelism. Image data is inherently parallel, and local neighborhoods have considerable overlap from one pixel to the next. Image algorithms are inherently parallel and are often implemented with long sequences of basic operations. Combined, this means hardware engineers have a rich design space with many degrees of freedom. In many ways, the high performance implementations of image processing algorithms that have been reported are due to this flexibility in design space. The hardware engineer can tailor the type and level or parallelization appropriate to a specific Reconfigurable Computer based on the number of gates and on-chip / off-chip memory bandwidth. Not many other applications have this luxury.

In this chapter we have described prototypical architectural solutions to several image processing problems. In practice, the details of these implementations for specific reconfigurable computing systems can greatly affect performance, and therefore, exploring the design space with the specific resource constraints is very important. Optimization under resource constraints is often what makes *hard*ware design *hard* for humans. Given the many degrees of freedom in image processing, optimal solutions are likely to remain *hard* for automated resource allocation tools and techniques as well. As computational capacity and memory bandwidth increase, we speculate that non-optimal, but sufficient solutions will become acceptable, simplifying the problem for both humans and machine.

7

Network Security

The need for secure communication over internetworks has increased greatly in the past decade. In today's networked society, it is necessary to protect information that passes through world-wide computer networks. It is necessary that data is read only by the intended recipient(s), that data is not compromised or altered during transmission, and that the recipient can be confident of the identity of the sender. It is necessary to make sure that unauthorized individuals do not gain access to private local area networks. Further, it is necessary to ensure that the network itself remains robust in the face of virus/worm attacks.

In this chapter, we discuss the role of reconfigurable computing to accelerate security-related communication. Applications include cryptography, network-based security protocols, and network protection.

7.1 Cryptographic Applications

Encryption ensures the secure communication of information among sender and recipients over insecure computer networks [357]. With the widespread use of public internetworking resources for financial transactions, electronic commerce, and distributed processing of confidential corporate data, bulk encryption of large data streams, especially via public key cryptography, has become increasingly important. The volume of data combined with the computational burden of encryption/decryption limit the utility of software cryptographic algorithms. E-commerce presents a burgeoning need for secure communication when the sender and receiver are not known to each other and raises issues of secure key exchange as well as identity authentication. Public key cryptography algorithms prove particularly computationally intensive, and map well to hardware.

In this section, we survey advances in reconfigurable hardware implementations of cryptographic algorithms.

7.1.1 Cryptography Basics

In cryptography, the original message, called *plaintext*, is given to a encryption algorithm along with a *key* to generate *ciphertext* (see Figure 7.1). The ciphertext travels over an insecure communication channel to the recipient, who applies the associated decryption algorithm using the same or a different key.

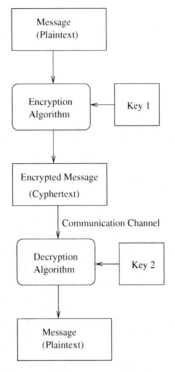

Fig. 7.1. A Cryptographic System

Symmetric Algorithms

When a Secret Key – also called symmetric – encryption algorithm is used, the encryption algorithm is public, but a single, shared key (Key 1 and Key 2 in the diagram are the same key) is kept secret between sender and recipient. Symmetric algorithms operate over data *blocks*, in which a group of bits (often 128 or larger) are encrypted as a unit; or over data *streams*, in which each bit in the stream is encrypted separately on the fly. Symmetric algorithms may also operate in a variety of *modes*. In feedback mode, the result of encrypting a block B_i is input to encrypt the next block B_{i+1}, thus requiring sequential computation of the list of blocks. In non-feedback mode , each block is encrypted independently, potentially in parallel.

Block Symmetric Algorithms typically consist of iterated "rounds," where the number of iterations depends on the size of the data block. A round consists of a sequence of operations. Figure 7.2 illustrates the structure of the Advanced Encryption Standard algorithm Rijndael [102].

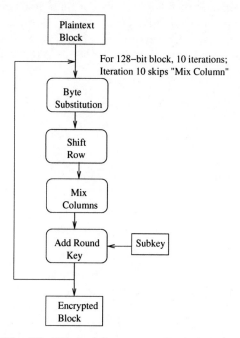

Fig. 7.2. Rijndael Symmetric Block Algorithm

In the Byte Substitution step, each byte is replaced by its substitute from a Substitution Table (S-Box). For each byte b, its S-Box entry is calculated by taking b^{-1} in $GF(2^8)$ and then applying an affine transformation over $GF(2)$ to b^{-1}. GF stands for Galois Field, a finite field over a prime number. In the Shift Rows step, rows of the block are shifted left, the shift amount depending on the row index. In Mix Columns, columns of the block are multiplied (in $GF(2^8)$) by a constant matrix. In the final Add Round Key step, a subkey derived from the original key that is unique to the round is XOR'ed with the block.

As this example shows, basic operations commonly used in block ciphers include

- logic operations such as XOR ,
- modular arithmetic in $GF(2^n)$ operations

- shifting
- permutation
- bit insert/extract
- table lookup

In addition to Rijndael, other well known block algorithms include 3DES, IDEA, and Blowfish.

Stream ciphers, in which a running stream of bits (or bytes) is encrypted one at a time, often depend on Linear Feedback Shift Registers (LFSR) in combination with simple logic circuits as shown in Figure 7.3. Taps on the LFSR are XOR'ed to generate a key stream, which is then XOR'ed the data stream to generate ciphertext. Popular stream ciphers are RC4, A5, and SEAL.

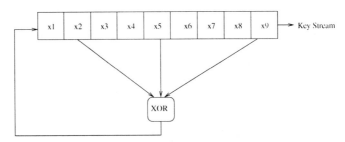

Fig. 7.3. Linear Feedback Shift Register

Asymmetric Algorithms

In Public Key – asymmetric – encryption algorithms, two keys are used, a public key Key 1 and a private key Key 2. Plaintext is encrypted using the public key, and decrypted using the private key. While the private key is derived from the public key, it is difficult to compute Key 2 from Key 1. This allows anyone to encrypt a message with the public Key 1, but only those holding private Key 2 can decrypt the message. Public key cryptography is based on the concept of one-way functions $F(k)$ in which the forward direction $F(k)$ is relatively simple to compute but the inverse direction $F^{-1}(k)$ is computationally intractable.

Certain public key algorithms can also be used for Digital Signature. A digital signature authenticates the identity of the sender of a message. To create a digital signature, the message is encrypted with the *private* key and transmitted. The recipient verifies the signature by decrypting with the public key. Public key algorithms are commonly used in combination with private key algorithms: a public key protocol is used to transmit the private key, and then the private key method is used during the session to exchange data. Using a combination of private and public key algorithms is beneficial because public key algorithms are significantly slower than private key. Popular public key algorithms include RSA, Diffie-Hellman, and Rabin.

A Message Digest ensures that a message is authentic and has not been altered during transmission. A one way hash function is used to generate a fixed size compressed representation of a message. To be used in a hash function, an arbitrary size message is divide into fixed size blocks. The hash algorithm iteratively hashes each block, using the previously hashed block as a second input (see Figure 7.4). An initialization vector is used for the first hash function application. The final output is the message digest. Most hash functions generate a 128-bit message digest from the arbitrary length message.

Like the private keys of asymmetric encryption, one way hash functions are easy to compute but difficult to reverse. In addition, given M, it is hard to find another message M' such that $H(M) = H(M')$. An additional property of collision resistance is also necessary. It must be difficult to find two messages M and M' with the same hash: $H(M) = H(M')$. Well known one way hash functions include MD5 , and the Secure Hash Algorithm (SHA). The latter is part of the Digital Signature Standard [306].

The message digest combined with digital signature is used to create a *Message Authentication Code (MAC)* as a tag appended to the message. One way hash functions are commonly used to generate MACs by concatenating the message with a shared, private key as a single input into the hash function. The digital signature protocol is then used: the message digest is encrypted with the private key and decrypted with the public key. This validates the authenticity of the sender as well as validity of the message.

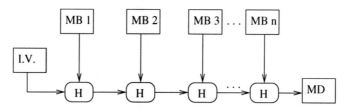

Fig. 7.4. Creating a Message Digest with a One Way Hash Function

The most common algebraic operations used to construct one-way functions for asymmetric algorithms are

- factoring. It is easy to multiply a pair of prime number $n = p \times q$, but given only n, it is difficult to find the prime factors p and q when p and q are big. To generate RSA keys, p and q large primes are chosen, with $n = p \times q$. Then e is chosen such that e is relatively prime[1] to $(p - 1) \times (q - 1)$. An

---

[1] Two integers are relatively prime if they share no common positive factors other than 1.

inverse of e, d, is also chosen ($e \times d = 1 \bmod (p - 1) \times (q - 1)$. Then n, e are published as the public key and d is the private key.

- calculating discrete logarithms and square roots modulo n. Given M, n, e, exponentiation in a finite field $C = M^e \bmod n$ is easy. However, knowing only C, it is difficult to recover $M = C^d \bmod n$. Calculation of discrete logs requires application of arithmetic operations in Galois Fields, finite fields over prime numbers p or their powers $GF(p)$, $GF(p^n)$. $GF(2^n)$, $(n > 0)$ is particularly attractive for computing in hardware.

- elliptic curve operations. Elliptic Curve Cryptosystems (ECC) have become increasingly attractive for generating public/private key pairs, especially in IPsec Internet security protocols (see Section 7.2). ECCs are similar to discrete log systems in that they utilize asymmetric operations in a finite field. Unlike discrete log systems whose elements are integers or polynomials, elements of ECCs are points in a 2-D Cartesian coordinate system that fall on a particular elliptic curve $y^2 = x^3 + ax + b$.

- hash operations. These include simple logic functions (AND, NOT, and XOR) over fixed length registers, and rotate/shift for fixed amounts.

7.1.2 RC Cryptographic Algorithm Implementations

Cryptographic algorithms are particularly well suited to reconfigurable logic implementation. These algorithms are compute-intensive, easily pipelined, and often use hardware-centric constructs, such as shift registers of various sizes and permutation networks. Modular arithmetic with arbitrary size of operands is also more efficiently implemented in hardware than with fixed width microprocessor ALUs. Reconfigurable logic hardware designs have been created for block and stream ciphers, public key algorithms, and one way hash functions.

A substantial body of research has been conducted on FPGA implementations of the algorithm chosen as AES standard, the Rijndael algorithm. The highest performance software implementations (in assembly language) have been reported at 580 Mb/s. Extremely high performance FPGA designs have been demonstrated. A fully pipelined AES encryption processor mapped to the Xilinx Virtex-II Pro is presented by [211] that achieves up to 21.5 Gb/s. The implementation unrolls all ten rounds. Implementation of the S-box function in logic using pipelined $GF(4)$ operations is compared to the use of on-chip Block RAM to look up the substitution bytes. A highly pipelined single-chip design that exploits on-chip Block RAM is reported to run at a data rate of 7–12 Gb/s in [297] and [284]. The UltraSONIC video processing system utilizes two Xilinx Virtex 1000E FPGAs to encrypt video at 2.1 Gb/s [297]. In contrast, [331] develop a compact design well suited to low end FPGAs with maximum throughput of 215 Mb/s.

The Serpent block cipher (an AES candidate) was implemented on FPGAs, and encryption rates greater that 4 Gb/s are shown to be feasible [136]. IDEA [74], DES [394] and 3DES have also been demonstrated on FPGAs.

Algorithm Type	Algorithm Name	FPGA	Area Slices	Memory BRAMs	Throughput
Block Sym pipelined	Rijndael AES	Virtex 1000	12600	0	12.2 Gb/s [163]
Block Sym pipelined	Rijndael AES	Virtex-II Pro	5177	84	21.54 Gb/s [211]
Block Sym iterative	Rijndael AES	Virtex 1000	2507		414.2 Mb/s [163]
Stream Sym	RC4	Virtex-II	138	3	176 Mb/s [239]
One Way Hash	SHA-256	Virtex V200	2120	0	326 Mb/s [371]
One Way Hash	MD5	Virtex 1000	4763	2	354 Mb/s [106]

Table 7.1. Representative Cryptographic Algorithms on FPGAs

There have been several FPGA designs based on the RC4 stream cipher. RC4 is used for encryption in wireless networking protocols. Throughput of 16–176 Mb/s using 9–19% of the available logic of a Xilinx Virtex-II have been reported in [194] and [239]. Results of FPGA implementations of one way hash algorithms (MD5, SHA-1, and SHA-256), as well as arithmetic functions within elliptic curve cryptosystems have been reported. Secure hash algorithms have also been implemented in reconfigurable logic, including SHA-2 [391], SHACAL (using SHA for block encryption) [285], and MD5 [106] with throughput in the range 326–480 Mb/s (SHA family) and 165–354 Mb/s (MD5).

RC designs for Elliptic Curve Cryptosystems have also been developed. [258] presents the design of a parameterized, microcoded elliptic curve processor in which the Arithmetic Logic Unit (ALU) performs operations in $GF(2^n)$ (using a normal basis) and the datapath is n bits wide. The elliptic curve operations of curve addition and multiplication are microcoded to use primitive $GF(2^n)$ operations. The processor was mapped to a Virtex V300. For ECC multiplication with $n = 281$, a speedup of 36X was achieved over a 270-MHz Sun Ultra-5. The $GF(2^n)$ normal basis (n=233) ECC design of [33] optimizes latency for a single data set rather than throughput for a larger data set. This is because the primary application of ECC operations is for digital signature applications. This work achieves speedup of 895–1300 over a Pentium 3. In [247], a linear array architecture is proposed for efficient multiplication in $GF(2^n)$ using a Gaussian normal basis. The work of [236] implements parameterized elliptic curve point multiplication over $GF(2^n)$ using polynomial basis.

Throughput results for representative FPGA implementations of cryptographic algorithms are summarized in Table 7.1.

Many of the algorithms developed recently exploit the enormous flexibility of FPGAs. Parameterized implementations allow the end user to instantiate designs optimized for specific block and key sizes, area vs. throughput, used of memory vs. logic. Area/delay trade-offs are considered in [437], while [308] compares the use of lookup tables versus direct logic implementation of a GF(256) multiplicative inverse. The effects of pipelining and other loop optimizations on AES implementations are considered in [163], [68] and [351].

It is even possible to partially, dynamically reconfigure some FPGAs to accommodate changing keys and modifying key and data block width. [286] presents methods to develop structured datapaths for an AES core using the Xilinx JBits tool, and then modify characteristics of the hardware circuit dynamically during execution, while [175] exploits partial reconfiguration in the IDEA algorithm by changing keys through dynamic reconfiguration.

7.2 Network Protocol Security

In this section we shift focus from security algorithms themselves to their use in secure computer network communication. Network communications is one of the earliest and largest market segments of commercial FPGA use. FPGAs are used in switches and routers in dedicated "ASIC-replacement" functions for high speed network backbones. The FPGAs are assembled into single function dedicated boards optimized for the specific switching or routing application.

Reconfigurable computing uses of FPGAs in networking combine low level network processing functions with application level processing in the network interface. The network interface can be used either at the router, a gateway, or at a workstation endpoint, with data rates ranging from 40+ Gb/s (for optical links on the backbone) to 1 Gb/s in local area networks. When a significant compute task such as encryption/decryption is combined in line with low level network processing, programmable hardware offers a high performance approach that can off-load the conventional microprocessor at data rates that exceed the capability of specialized network processors.

In this section we discuss network security applications that use reconfigurable computing. We first describe network interface cards that include RC processors, and then discuss two broad application areas – in-line encryption/decryption for secure packet transport and signature-based network intrusion detection. Both application domains are well suited to reconfigurable computing, combining low level packet assembly and transmission with compute-intensive processing of the data stream.

7.2.1 RC Network Interface

A generic RC network interface is shown in Figure 7.5. There are two parts, the physical layer interface and the reconfigurable computer.

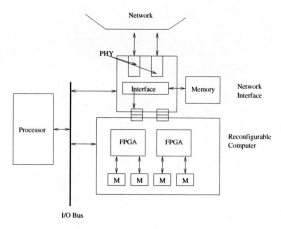

Fig. 7.5. A Generic RC Network Interface

The network interface component contains ASICs that communicate with the physical network (labeled PHY in the Figure). Glue interface logic, often mapped onto a dedicated fixed function FPGA, assembles packets and pushes them onto FIFO queues. Often a dedicated memory is used to buffer packets. Network packets may flow directly to a host processor over the I/O bus. Alternatively, the reconfigurable computer can read the FIFO queues and perform application-specific processing. Processed packets then can either go to the host processor or back out onto the network via output FIFOs.

Examples of RC network interfaces include a firewall inline processor (FIP) [283], the Field-programmable Port Extender (FPX) [57], and the Gigabit Rate IPsec (GRIP) card [41].

The firewall inline processor interfaces to ATM networks. SONET/SDH framing chips provide an interface between the ATM physical line interface and the digital control and processing on the RC. There are three ports to the network. Physical transmission of data to and from the FIP is accomplished by HFBR-5205 optical transceiver pairs. As mentioned above, on-board SRAM is used to buffer packets at the network interface. The reconfigurable computer, an ORCA FPGA, filters ATM packets, sending control information to the firewall host, while at the same time allowing friendly connections to proceed without performance degradation. The algorithms are customized to the ATM format and protocol. The firewall processor, built in the mid-1990's was able to keep up with a 155 Mb/s transmission speed of the ATM link, processing a 53-byte ATM cell every 2.74 microseconds.

Several in-line filtering applications have been demonstrated on the Field-programmable Port Extender, a circa-2000 implementation of a reconfigurable computing network interface used in a router or gateway.

The network interface part of the FPX (see Figure 7.6 uses a Xilinx Virtex XCV600E to interface to two multi-Gb/s line cards (Gigabit Ethernet and

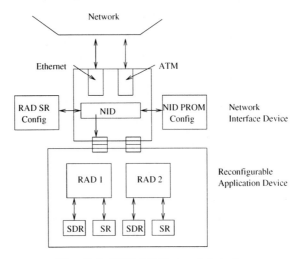

Fig. 7.6. FPX RC Network Interface

ATM). It also controls a configuration SRAM to load new packet processing algorithms into the reconfigurable computer. Configurations can be received over the network, allowing remote reconfiguration of the RC. The RC is a Xilinx Virtex XCV2000E with access to off-chip SDRAM and SRAM banks. Applications that have been implemented on the FPX include internet header filtering using exact match or longest prefix matching. Regular expression matching within the payload has also been implemented as well as a Bloom filter to scan a large number of fixed-length strings in hardware. Protocol wrappers have been developed to parse Internet Protocol (IP) and Transmission Control Protocol (TCP) format flows.

In contrast to the FPX, the GRIP network processing accelerator card is a host-based RC network interface. Like the FPX, it uses Xilinx Virtex FP-GAs both for network interface function as well as reconfigurable computer. GRIP integrates into a standard Linux implementation of the TCP/IP/IPsec protocols. It provides full-duplex gigabit-rate acceleration of a variety of cryptographic algorithms operations and other application-specific kernels.

The overall GRIP system is diagrammed in Figure 7.7.

Host software to communicate with GRIP consists of a high-performance device driver and special interactions with the operating system. The network interface portion of GRIP is a custom Gigabit Ethernet mezzanine daughter card on the reconfigurable computer, a SLAAC-1V FPGA co-processor board [39]. The network interface daughter card has a Vitesse 8840 Media Access Controller (MAC), and a Xilinx Virtex 300 FPGA which interfaces to the X0 chip through a 72-bit connector. The Virtex 300 uses 1 MB of external ZBT-SRAM for packet buffering, and performs common off-load functions such as filtering and IP checksumming.

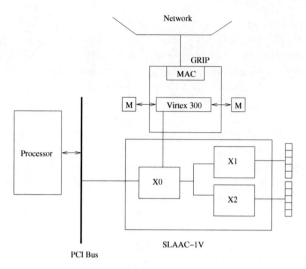

Fig. 7.7. GRIP RC Network Interface

The X0 device of the SLAAC-1V acts as a dedicated packet mover and PCI interface, The remaining two devices (X1 and X2) act as independent transmit and receive processing pipelines, and can be loaded with arbitrary computational modules. A number of packet-processing cores   have been developed on the GRIP platform, including AES (Rijndael), 3DES, SHA-1, SHA-512, snort-based traffic analysis, rules-based packet filtering (firewall), and intrusion detection.

7.2.2 Security Protocols

Ensuring secure communications over insecure public computer networks can be provided at several different layers of the network stack. At the data link layer, the lowest level network packets are encrypted as they leave a node and decrypted as they are received. Since all bits – including headers – are encrypted, decryption and encryption must be done at every intermediate stage (router, gateway) of the network. Data link encryption is most suitable for dedicated high value networks. At the TCP level, the Secure Socket Layer [84] has emerged as a standard for secure web communication. SSL uses asymmetric public key algorithms to establish a secure connection and then symmetric algorithms such as 3DES and RC4 within a session.

The IPsec Internet Protocol Security standard [1] specifies encryption at the IP layer of the network stack. IPsec is neutral to the encryption algorithm. In fact a different algorithm may be negotiated for every session for both private and public key applications. This requirement for flexibility has made IPsec an attractive protocol to accelerate with reconfigurable logic. Using RC, it is possible to switch among encryption algorithms using the same programmable hardware. It is possible to update algorithms in the field by dynamically reconfiguring the hardware during operation to specialize the hardware circuit, e.g., to a particular key [321]. It is possible for some RC implementations to exceed ASICs in performance. The work of [103] designs an adaptive cryptographic architecture for IPsec in which hardware crypto-graphic algorithms can be swapped into the FPGA on demand.

It should be noted that overall system performance is a function of many factors, and that protocol overhead can reduce the benefit of the reconfigurable computer. In [73], a virtual private network implementation using FreeS/WAN is accelerated using the Pilchard reconfigurable computer. Pilchard communi-cates with a host over a 64-bit DRAM interface, providing high bandwidth, low latency communication between processor and a Virtex 1000E. The Pilchard FPGA is configured with a 3DES core in feedback mode. The maximum throughput of 136 Mb/s is 3X software speed. However, since encryption ac-counts for only 50% of the total compute time for the VPN, by Amdhal's law[2], the maximum speedup of the overall application is only 50%, not 300%.

The GRIP project incorporates IPsec processing into the network inter-face card, enabling complex cryptographic transformation with IPsec to be executed in hardware. The driving application of multimedia packets (such as HDTV [40]) over the Internet differs from VPN in that almost all of the pack-ets must be encrypted and decrypted – handshake and protocol exchange are insignificant fractions of the overall communication – and thus the hardware speedup is better reflected in overall system performance. From the host, raw IP/IPsec packets are passed to the GRIP software driver with all the appropri-ate header information but no encryption. The GRIP driver looks up security parameters (key, Initialization Vector , algorithm, etc.) for the corresponding IPsec session, and prefixes these parameters to each packet before handing it off to the hardware. The X0 device fetches the packet across the PCI bus and passes it to the transmit pipeline (X1). X1 analyzes the packet headers and security prefix, encrypting or providing other security services as specified by the driver. The packet, now completed, is sent to the Ethernet interface on the daughter card. The receive pipeline is just the inverse, passing through the X2 FPGA for decryption. Measured performance of AES-encrypted HDTV packets transported over IP at 900 MB/s is reported.

7.2.3 Network Defense

With programmable logic available right at the network interface, it is possible to screen packets as they enter a host, a gateway , or router/switch (see Figure 7.8). For example, at a gateway, it is common to use network intrusion detection software and examine each packet entering the Local Area Network. The network administrator prepares a rule database. A rule consists of a

---

[2] Maximum speedup $S = 1/[\alpha + (1-\alpha)/P]$ where α is the sequential fraction and P is the hardware speed.

condition and an action. The condition is specified as a logical expression
whose clauses are regular expressions matching fields in the IP header or in
the packet body. If the logical expression when applied to a packet evaluates
to true, the action is performed.

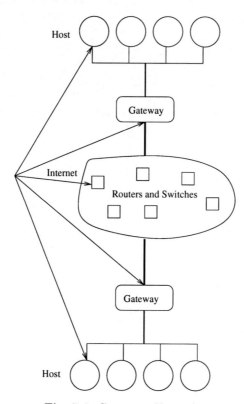

Fig. 7.8. Computer Networks

Researchers have exploited this capability by designing pattern match
hardware circuits to scan network packets for viruses and worms, as shown
in Figure 7.9. Using the GRIP network interface to acquire Ethernet packets, the Gigabit Rate Network Intrusion Detection Technology (GRANIDT)
project [172] configured X1 and X2 of the SLAAC-1V for header and packet
content filtering. A modified "snort" [345] software runs on the host. It parses
each rule, which consists of conjunction, disjunction, or negation of pattern
specification clauses and creates the filter database. Clauses specifying patterns in headers are processed in X1 while content matching regular expressions are processed by X2. Part of the FPGA is configured as a tertiary
Content-Addressable Memory (CAM), so that new clauses can be loaded without having to reconfigure the device. The clause results are returned to the
host as a bit vector, and software combines the clauses according to the rule

database to determine which rule (if any) a packet matches. Exploiting the FPGA's spatial parallelism, it is possible to match many clauses simultaneously, and thus keep up with multi-gigabit rate traffic, which is not possible in software. Another approach to pattern matching using CAMs is reported in [374].

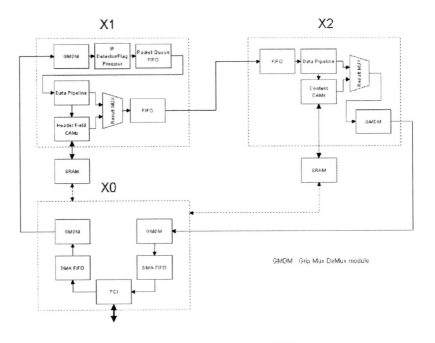

Fig. 7.9. Packet Scanning on GRIP

Many methods for converting regular expressions into hardware for network packet filtering and intrusion detection applications have been proposed in the reconfigurable computing literature. Methods for generating hardware to match specific regular expressions are presented in [222], [299], [255], [76], [34], and [83]. Since there are hundreds to thousands of regular expressions to match, optimizations are proposed to share logic to recognize sub-expressions. The drawback of these approaches is that new hardware must be synthesized each time the rule database changes, which is not required by the CAM method.

In contrast to direct hardware implementation of regular expressions, [287] designs a network processing accelerator consisting of a modified 3-level barrel shifter network that processes data from a 32-byte or 128-byte input buffer. Tree lookup (useful for routing) and pattern match algorithms are presented in the form of sequences of instructions to the accelerator. Another alternative employing Bloom filters is proposed by [115].

7.3 Summary

We have seen that reconfigurable computing plays many different roles in secure network communications. Using hardware-friendly primitives such as simple logic operation, modular arithmetic, and shift/rotate operations, a large body of encryption algorithms have been mapped onto FPGAs. Speedup of 2X–10X over software implementations have been reported. The ability to place these programmable hardware encrypter/decrypter in line with the network interface makes it possible to off-load compute- and data-intensive functions from the control processor.

Network interface cards containing reconfigurable hardware have been used for a variety of network security applications. By scanning packets as they arrive in real-time, RC modules can match packet headers and content for arbitrary patterns. Patterns to be matched can be updated over time, allowing the network interface to respond to new threats. Regular expression pattern match of header fields enables the network interface to accelerate firewall enforcement. When the reconfigurable hardware is used to scan packet (or message) content, the network interface can detect worms or viruses in real time.

Bioinformatics Applications
Dominique Lavenier and Mathieu Giraud

8.1 Introduction

Bioinformatics refers to the analysis and the management of biological information. The term computational biology is more often used to address physical and mathematical simulations of biological processes. The need for bioinformatics capabilities has been precipitated by the explosion of publicly available genomic information resulting from the Human Genome Project. The goal was to determine the sequence of the entire human genome (approximately three billion characters named *base pairs, bp*). The science of bioinformatics, which is the melding of molecular biology with computer science, is essential to the use of genomic information in understanding human diseases and in the identification of new molecular targets for drug discovery.

Bioinformatics covers a large field of domains ranging from text sequence analysis to gene network modeling, 3D protein structure prediction, gene expression data analysis through DNA-chip, phylogeny, etc. But, for the last decade, the deluge of genomic information has led to an absolute requirement for computerized databases to store, organize, and index the data and for specialized tools to view and analyze genomic data. It is in this last domain – analysis of genomic data – that FPGA accelerators have exhibited extremely good performance for time consuming bioinformatics computations over a huge volume of data. They have demonstrated that they can be a viable alternative compared to supercomputers or clusters in both industrial [441,443] and academic projects dedicated to intensive bio-computing algorithms : in 2003 and 2004, more than ten different teams published papers on FPGA implementations of sequence similarity search!

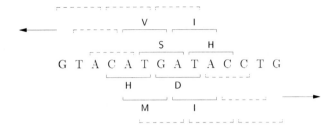

Fig. 8.1. Translation of a DNA string to protein strings according to the 6 different reading frames.

Genomic data consist of DNA or protein sequences. A DNA sequence may be as simple as a single gene (a few thousands characters) or as complex as a complete genome (three billions characters for the human genome). DNA strings are sequences of *nucleotides* that range over the 4-letters alphabet $\Sigma_4 = \{A, C, G, T\}$. This alphabet leads to a compact 2-bit encoding scheme, well suited for FPGA hardware optimization compared to a 32- or 64-bit Von Neumann architecture. Protein sequences are shorter. They are *translated* from DNA genes on 6 possible *reading frames* (Figure 8.1). Their length range from a few hundred to a few thousands characters over a 20-letter alphabet of *amino acids* $\Sigma_{20} = \{A, C, D, E, F, G, H, I, K, L, M, N, P, Q, R, S, T, V, W, Y\}$.

DNA and protein sequences are stored in genomic databases. SWISS-PROT and TREMBL [144], for example, are two well-known protein sequence databases containing respectively 166,000 and 1,154,000 sequences (Dec. 2004). DNA databases GenBank and EMBL (Genbank release 145, Dec. 2004) contain more than 40 million sequences, representing 44 billion nucleotides [46, 146]. GenBank and EMBL are part of the International Nucleotide Sequence Database Collaboration, which also comprises the DNA Data Bank of Japan (DDBJ). These three organizations exchange data on a daily basis and new releases are made every two months to include new data coming from worldwide research institutes. As these banks are growing exponentially (every year the number of sequences nearly doubles), performing computation on the growing mass of data is more and more challenging.

Most genomic computations rely on making comparisons over these sequences, either for highlighting common sub-sequences, computing edit distances or finding similarities. Many bioinformatic algorithms require approximate string and pattern matching. The arithmetic is mostly integer operations on small data width operands. Floating point operation is generally not required. Furthermore, many fine grained parallel implementations of string comparison algorithms have been proposed and successfully tested on dedicated architecture. In this context, FPGA-based machines are very well suited to support bioinformatics algorithms related to DNA or protein sequence processing.

The next section presents bioinformatics applications for which reconfigurable computing is a good candidate. Section 8.3 presents the implementation of the fundamental sequence comparison algorithm based on dynamic programing. Section 8.4 is dedicated to BLAST-like seed-based heuristics. Section 8.5 deals with profiles, stochastic models, and languages. Section 8.6 concludes the chapter by describing some reconfigurable accelerators efficiently supporting bioinformatics sequence processing algorithms.

8.2 Applications

In this section, we detail various applications related to genomic computations which can benefit from FPGA technology. As the primary data sources are DNA or protein sequences, string comparison algorithms are of great interest. This survey is by no means exhaustive. We give some representative examples and highlight the need for computing power.

8.2.1 Genome Assembly

Before analyzing genomes in-silico, it is necessary to sequence the long DNA molecule contained in each cell of every living organism. The most successful method to accomplish DNA sequencing is to break the DNA molecule into millions of random short fragments which are then re-assembled to reconstruct the final text of the genome. This method, called *shotgun sequencing* , was introduced by Fred Sanger in 1982 [145] and is now used to sequence large genomes such as mouse or human genomes.

Many algorithms have been proposed to rebuild text of genomes from these short random DNA fragments [328]. A common pre-processing step for *overlap-layout-consensus* algorithms is to find overlapping regions between fragments, which requires making exhaustive pairwise comparison to detect similarity between the beginnings and the ends of all fragments. In other words, assembling N fragments lead to $O(N^2)$ pairwise fragment comparisons.

The quality of the sequencing depends on fragment oversampling: the higher the number of fragments, the higher the probability of high quality sequencing. The term *coverage* refers to the number of fragments that overlap a particular position in the sequence. As an example, sequencing a genome of length $M = 3 \times 10^9$ *bp* (mouse) with fragments of size $K = 600$ *bp* and a coverage $C = 5$ leads to $N = C \times M/K = 2.5 \times 10^7$ fragments. More than 3×10^{14} elementary pairwise comparisons must be performed to assemble the mouse DNA genome .

Algorithms to search textual similarities between two fragments can be highly parallelized and efficiently mapped onto reconfigurable computing machines, resulting in faster pre-processing.

8.2.2 Content-Based Search

Genomic banks are routinely and daily scrutinized by thousands of researchers. A common task in molecular biology is to try to assign a function to an unknown gene or an unknown protein. More precisely, proteins are synthesized within the cells of plants and animals. To function, a protein must adopt a specific 3D shape related to its sequence of amino acids. The shape is important because it helps determine the function of the protein and how the protein interacts with other molecules. It is assumed that two proteins with identical functions may have similar 3D structures, and thus, a similar sequence of amino acids. Even if this hypothesis is not always verified, a great number of algorithms have been proposed to rapidly extract sequences (or portions of sequences) having a high similarity with a query sequence.

When scanning genomic databases, biologists are faced with a dilemma: speed or quality. Genomic data grow exponentially ($\times$ 2 every $12-15$ months) and at a faster rate than computing power ($\times 2$ every 18 months). On a standard computer, a high quality search may take hours while a more approximate search can just be performed in a few tens of seconds. High-quality searches are based on time consuming dynamic programming methods (Section 8.3). Other search algorithms, such as the well-known BLAST program, are based on seed-based heuristics that can dramatically reduce computing time (Section 8.4).

Reconfigurable computing machines can help to speed up both types of algorithms. Parallel implementation of dynamic programming algorithms for genomic computation on VLSI chips or FPGA components have demonstrated that speedup of at least two orders of magnitude can be achieved. In the same way, hardware implementations of seed-based heuristics on reconfigurable computers can also decrease the computation time. This is particularly important for BLAST servers which have to process millions of requests every day.

8.2.3 Genome Comparison

By the end of 2004, about 240 genomes had been completely sequenced, and about 1000 other genome sequencing projects were underway [442]. In comparison, there were 650 projects referenced in December 2002, and approximately 350 in December 2000. There is no reason to expect any decline in the next few years. More and more complete genomes will be sequenced, coming from a large diversity of living organisms: virus, bacterium, plants, fishes, vertebrates, etc.

This opens the door to new ways of investigating the genome structures. The full genomic sequences of different species can be compared in order to highlight conserved regions, duplicated or repeated zones, chromosome rearrangement, etc. Hence, from gene level (sequence of thousands of characters) we now move to genome level (sequence of hundred millions of characters) analysis.

From a computational point of view, genome comparison algorithms do not differ from standard string comparison algorithms, except that the length of the sequences may limit their use. Strings of hundred of millions of characters must be processed to detect regions of interesting similarities. This is a very time-consuming task which can be greatly accelerated by FPGA-based reconfigurable computing machines.

Compared to gene analysis, which can be satisfactorily performed (in time) on a standard computer, genome analysis increases the complexity by several orders of magnitude. Generally, coarse grained parallelism through supercomputers, large clusters of PC or grid computing allows biologists to run large-scale comparison on whole genomes. FPGA-based machines best exploit fine grained parallelism and can be tailored to the requirements of DNA or protein data structures.

8.2.4 Molecular Phylogeny

Biologists estimate that there exist millions of different species of living organisms on the Earth. Morphological criteria and gene structure suggest that they are genetically related, and their genealogical relationships can be represented by a vast evolutionary tree. This tree represents the phylogeny of organisms, that is the history of organism lineages as they change through time. It implies that different species arise from previous forms via descent, and that all organisms are connected by the passage of genes along the branches of the phylogenetic tree.

To build the phylogenetic tree of a group of organisms, identical (or near identical) genes present in all them are first systematically compared. This aims to calculate a *distance* between all pairs of genes. The greater the distance, the older the relationship between genes: it is assumed that genes mutate independently after speciation. Based on this matrix of distances, trees can be constructed through different phylogenetic methods.

Again, the pre-process step involves string comparison to compute the matrix distance. Furthermore, some methods need to periodically re-evaluate the matrix during the tree building process. While reconfigurable computers are mainly used to compute gene distances, efforts have also been made to speed up the whole process on Xilinx Virtex FPGAs, especially using the Maximum Likelihood approach [275, 277].

8.2.5 Pattern Matching

Comparing several proteins to regroup them in functional families is based on the similarity of their sequences. Most of the time, proteins from a same family indeed come from a common ancestor and contain *domains* of several to tens of amino acids that were conserved during evolution. Once those regions of similarity have been found, one can characterize these domains with patterns. These patterns often represent active sites in the protein and can be considered as a signature related to a specific function. As an example, the dog olfactory receptors constitute a family containing more than 1100 genes sharing up to five different specific patterns.

A pattern can be an exact word or alternatively a finite dictionary as in the PROSITE pattern database [220] in which known families of proteins are grouped. An example of this syntax is the pattern L-x(1,3)-M-x-[FILY]-D-R where [FILY] is a choice between four amino acids (F, I, L and Y) and x(1,3) a gap (of any amino acids) whose length is between 1 and 3.

Two types of computation can be considered: searching a pattern database with a query, or searching a protein or a DNA database with a pattern. The second case can be a very time-consuming task, especially when DNA database are scanned with a protein pattern. In that case, the database need to be searched into the six reading frames (Figure 8.1). FPGA can efficiently parallelize this task through direct automaton implementations.

8.2.6 Protein Domain Databases

Analysis of the fast growing number of full text genomes confirms that organisms as diverse as bacteria and human share many proteins and proteins domains. The total number of different folds that protein modules could adopt is estimated to range between 1000 to 6000 [58]. One of the challenge of the post-genomic era is to systematically characterize the repertory of protein domains and their interaction in terms of biological function [413]. Since proteins can be composed of one or more domains, methods for both determining these domains and parsing the protein domain databases are of great interest.

Among the various methods to identify protein domains, homology-based methods, such as MKDOM   [178] or DOMAINATION   [166], attempt to find multiple alignments. Methods based on HMM profiles, such as HMMER [135], have also been developed. Data from the leading protein databases (SWISS-PROT, TREMBL [144]) are used to automatically process millions of amino acid sequences and generate specific protein domain databases such as PRODOM   [95] or PFAM [143].

Some algorithms involved in multiple sequence alignments or HMM profiles are iteratively searched : the PSI-BLAST [148] program used in MKDOM, for example, first searches the protein database with a simple query. Then, from the results(a set of sequences), a profile is made, and the database is again queried. This process is iterated several times: each iteration enhances the profile with new sequences. Ideally, the process stops when no new sequences are detected. This must be repeated billions of times to generate the PRODOM database.

8.3 Dynamic Programming Algorithms

This section describes string comparison algorithms based on dynamic programming (DP) methods and their hardware implementation. DP methods were first proposed by Needleman and Wunsch (NW [309]) and Smith and Waterman (SW [373]) in 1970 and 1981 respectively. The NW algorithm evaluates the similarity between two DNA or protein sequences (global alignment), while SW finds two high similarity sub-sequences (local alignment) . Both compute an *alignment cost* that can be viewed as a *similarity score*.

8.3.1 Alignments

Genomic sequences are strings over the nucleotide alphabet Σ_4 or the proteic alphabet Σ_{20}. However, some applications use IUPAC ambiguity codes that allow 15 combinations of one or more nucleotides. Measuring the similarity between two strings means to try to *align* them to find their similarities, and to estimate the *cost* of transforming one string, character by character, to the other. At a given position, one of three cases can occur (Figure 8.2):

- a *match* occurs when the same character α is present in both strings,
- a *mismatch*, also called a *substitution*, when there are two different characters α and β,
- and a *gap*, when there is an *insertion* of one character in only one string, or symmetrically a *deletion* in the other string.

$$
\begin{array}{c}
\text{T T G A A A T G C G} - \text{A G T} \\
|\ |\quad\ |\quad |\ |\quad\ |\ |\quad\ |\ |\ | \\
\text{T T C A T A T} - \text{C G T A G T}
\end{array}
$$

Fig. 8.2. Global alignment of two DNA strings with 10 matches, two mismatches G/C and A/T, and two gaps G/– and –/T.

When dealing with amino acids (protein sequence), matches or mismatches can be more or less significant. A *score* is thus assigned to each pair of amino acids through a substitution function $d(\alpha, \beta)$ measuring the similarity between α and β. For example, the BLOSUM62 matrix refers to the substitution probability of amino acids [206], taking into account the evolution of living organisms over hundred of generations (Figure 8.3). In a similar way, one can assign a penalty $g_{penalty}$ for each gap that can be viewed as special cases of the function $d(\alpha, -) = g_{penalty}$ or $d(-, \beta) = g_{penalty}$.

	D	E	H	I	L
D	6				
E	2	5			
H	-1	0	8		
I	-3	-3	-3	4	
L	-4	-3	-3	2	4

Fig. 8.3. Extract of the BLOSUM62 matrix that assigns to every pair of amino acids a substitution score [206].

8.3.2 Dynamic Programming Equations

More formally, let $X = (x_1, x_2 \ldots x_m)$ and $Y = (y_1, y_2 \ldots y_n)$ be two strings to be compared, and $H(i, j)$ the maximum similarity score of the two subsequences $x_1 \ldots x_i$ and $y_1 \ldots y_j$. $H(i, j)$ can be computed in a two dimensional recursive form using the Needleman-Wunsch equation (NW, [309]):

$$\forall i: \quad H(i, 0) = g_{\text{penalty}} \times i \qquad \forall j: \quad H(0, j) = g_{\text{penalty}} \times j$$

$$\forall i, j, ij \neq 0:$$

$$H(i, j) = \max \begin{cases} H(i - 1, j - 1) + d(x_i, y_j) & \text{(match or substitution)} \\ H(i - 1, j) - g_{\text{penalty}} & \text{(insertion)} \\ H(i, j - 1) - g_{\text{penalty}} & \text{(deletion)} \end{cases} \tag{8.1}$$

An example of a NW computation is given in Figure 8.4a. The final quantity $H(m, n)$ is the *global similarity* between X and Y. In many applications, it is desirable to study *local similarities* and find the most similar sub-sequences of X and Y. Now $H(i, j)$ is the similarity score between the most similar pair of sub-sequences ending at x_i and at y_j. This leads to the Smith-Waterman equation (SW, [373]):

$$\forall i, j: \quad H(i, 0) = H(0, j) = 0$$

$$\forall i, j, ij \neq 0:$$

$$H(i, j) = \max \begin{cases} 0 & \text{(local align. starts here)} \\ H(i - 1, j - 1) + d(x_i, y_j) & \text{(match or substitution)} \\ H(i - 1, j) - g_{\text{penalty}} & \text{(insertion)} \\ H(i, j - 1) - g_{\text{penalty}} & \text{(deletion)} \end{cases} \tag{8.2}$$

An example of a SW computation is given on Figure 8.4b. In both NW and SW equations, the score propagation through the DP matrix is purely local: each (i, j) inner cell receives previous scores from three neighbors and

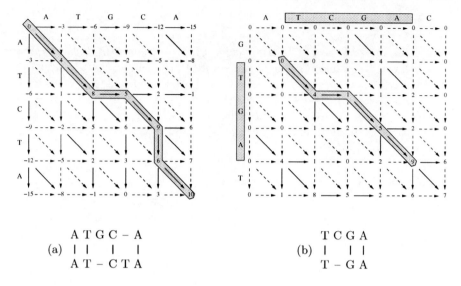

$$
\begin{array}{ccccc}
\text{A T G C} & - & \text{A} \\
\text{(a)} \quad | \ | & & | & | \\
\text{A T} & - & \text{C T A}
\end{array}
\qquad
\begin{array}{ccc}
\text{T C G A} \\
\text{(b)} \quad | & | \ | \\
\text{T} & - & \text{G A}
\end{array}
$$

Fig. 8.4. Example of global alignment computation with the NW equation (a) and local alignment computation with the SW equation (b). Scores are $+4$ for a match, -2 for a mismatch, and -3 for a gap. The solid arrows are the dependencies that lead to the maximum of each (i, j) cell and reveal the best alignment.

sends its results to three other cells (Figure 8.5a). The total number of cells is $O(mn)$, but the m cells on a same anti-diagonal can be computed in the same time (Figure 8.5b).

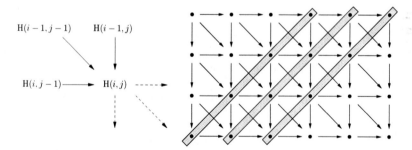

Fig. 8.5. Locality of the computation of a NW/SW cell. The solid arrows are the scores coming from previous cells. The only other quantity the cell needs is $d(x_i, y_j)$. Dashed arrows show the data propagation for the cells following (i, j). When computing every cell of the DP matrix, all the cells on a same anti-diagonal can be computed simultaneously.

8.3.3 Gap Functions

In equations (8.1) and (8.2), the penalty $g_{penalty}$ is a constant. It can be viewed as a linear function $g(\ell) = g_{penalty}\,\ell$ involving the length ℓ of the actual gap. More significant gap functions reflecting a better biological reality can be considered: it is often more costly to open a new gap than to extend an existing one. A commonly used function is the affine function $g(\ell) = g_{open} + \ell\,g_{extend}$. In this case, the recursive part of the DP equations can use several matrices, as shown by Gotoh in 1982 [177]:

$$\forall i, j, ij \neq 0 :$$

$$H(i,j) = \max \begin{cases} 0 & \text{(if SW local alignment)} \\ H(i-1,j-1) + d(x_i, y_j) & \text{(match or substitution)} \\ I(i,j) & \text{(insertion)} \\ D(i,j) & \text{(deletion)} \end{cases} \tag{8.3}$$

with:
$$I(i,j) = \max \begin{cases} H(i-1,j) - g_{open} \\ I(i-1,j) - g_{extend} \end{cases} \qquad D(i,j) = \max \begin{cases} H(i,j-1) - g_{open} \\ D(i,j-1) - g_{extend} \end{cases}$$

With such an affine gap function, the number of operations needed to compute all cells remains $O(mn)$ [177].

8.3.4 Systolic DP Computation

Following the methodology of Kung [246], the DP array can be projected on a systolic array (Figure 8.6). Lipton and Lopresti proposed in 1985 a bidirectional systolic array [266] in which the two strings propagate in opposite directions (Figure 8.6a).

Unidirectional systolic arrays dedicated to genomic computation were proposed in 1991 by Chow [80] and in 1993 by Hoang [210]. One string is loaded and stored in the systolic array, then the other is processed (Figure 8.6b). A minimal number of $\min(m + 1, n + 1)$ systolic cells are needed and the computation is done in $O(m + n)$ cycles. Figure 8.8 shows the operation of a unidirectional systolic array. Eleven time steps are illustrated. The shaded blocks indicate times when cells are inactive. For example, in the second time step (row 2) cells holding y_3, y_4, and y_5 are not active.

The $\min(m + 1, n + 1)$ systolic cells of the unidirectional can be compared to the $m + n - 1$ cells needed by the bidirectional array: in general, the unidirectional array is more efficient. Nevertheless, the bidirectional array can be more suitable for processing two large strings for a global alignment while staying close to the main diagonal (Figure 8.7).

Both unidirectional and bidirectional systolic arrays utilize simple, locally connected systolic cells, as illustrated in Figure 8.9.

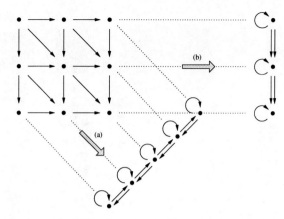

Fig. 8.6. Projections of a DP array to systolic cells. The diagonal projection (a) leads to a bidirectional array. The horizontal projection (b) leads to a unidirectional array.

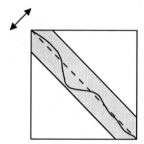

Fig. 8.7. A few cells in a bidirectional systolic array can be enough to globally align large sequences if we constrain the computations to a small zone around the main diagonal.

8.3.5 Backtracking

The NW/SW equations give the best global or local similarity score. Although it is sufficient in most applications, for instance when filtering a large database with short queries, the actual problem is to reveal the alignment that produced the best score.

In this case, backtracking information must be kept. In other words, each cell in the DP matrix needs to remember where its score comes from. This information, showed by the solid arrows on Figure 8.4, is one of the values ←, ↖, or ↑. The systolic cell can store the information in a local LIFO stack. Then a backtracking phase follows the score computation phase to build the alignment [210].

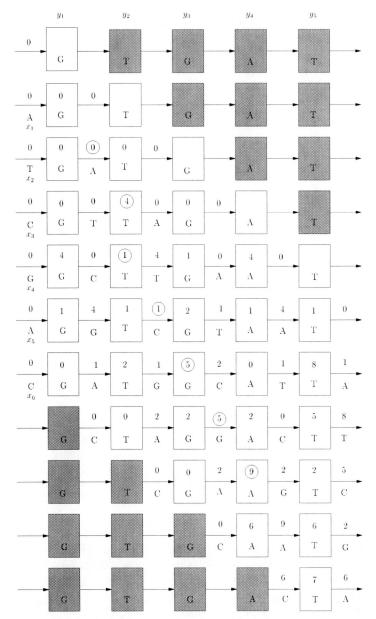

Fig. 8.8. Run of a unidirectional systolic array similar as those of Hoang [210] matching X = ATCGAC against Y = GTGAT. This run computes the SW matrix described on Figure 8.4b. The string Y is supposed to be loaded before the start of the computation. The circled values are those of the optimal local alignment (the best score is 9).

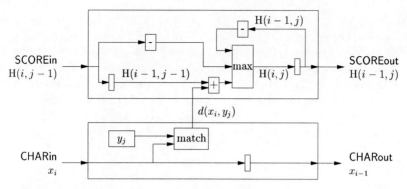

Fig. 8.9. Detail of a NW systolic cell. For a SW systolic cell, the maximum operator receives an additional input 0. The DP equations need only the value $d(x_i, y_j)$, and not the characters x_i and y_j themselves. Therefore the matching phase (computation of $d(x_i, y_j)$) and the DP phase (recursive computation of $H(i, j)$) can be separated [96]. Additional control logic, not shown here, manages the initialization phase (storing CHARin in the memory y_j) and the final phase (with optional backtracking).

8.3.6 Modulo Encoding

There are many different sorts of biological scores, but one, the $-1/-2$ *edit-distance score*, is of particular significance:

$$\begin{cases} g_{\text{penalty}} = -1 \\ \forall \alpha, \, d(\alpha, \alpha) = 0 \\ \forall \alpha, \beta, \, \alpha \neq \beta, \, d(\alpha, \beta) = -2 \end{cases}$$

This equation computes the minimum number of insertion and deletion operations between two strings: a substitution counts for one insertion plus one deletion. The edit-distance score has a very interesting property established by Lipton and Lopresti in 1985 [266]:

$$H(i, j) = H(i - 1, j) \pm 1 \quad \text{and} \quad H(i, j) = H(i, j - 1) \pm 1$$

Thus $H(i, j) - H(i - 1, j - 1) \in \{-2, 0\}$. Because of this, the general NW equation (8.1) can be rewritten as:

$$\forall i, j, ij \neq 0:$$

$$H(i, j) = \begin{vmatrix} H(i - 1, j - 1) & \text{if } H(i - 1, j) = H(i - 1, j - 1) + 1 \\ & \text{or } H(i, j - 1) = H(i - 1, j - 1) + 1 \\ & \text{or } (x_i = y_j) \\ H(i - 1, j - 1) - 2 & \text{elsewhere} \end{vmatrix} \tag{8.4}$$

Thus the value in the DP matrix can be represented modulo 4 with only two bits. Since modular arithmetic is used, overflow never occurs. This modulo

encoding was found by Lipton and Lopresti in 1985 [266] and can be extended
for amino acids scores [210]. For a global NW alignment, such an assumption
reduces the problem to the longest common subsequence (LCS) problem that
can be accelerated in software with bit-vector operations [98, 132].

As stated by Dydel, recent FPGA implementations of the edit distance
score with modulo encoding should be compared to such optimized software
techniques [132]. More generally, although the modulo encoding implementation worked well on early generations of FPGAs when the available logic
resources were more limited, modulo encoding can be overly restrictive: most
of the time, biologists prefer more realistic scoring schemes.

8.3.7 FPGA Implementations

DP algorithms were first accelerated by the P-NAC systolic array, an ASIC
solution using the *modulo encoding* scheme proposed by Lopresti [270]. The
first FPGA implementations were on the Splash and Splash 2 FPGA systolic arrays [168, 210, 271] in the beginning of the 90's. More recently, other
implementations have been proposed by the HokieGene team [333] and Yu,
Kwong, Lee and Leong [434]. Whereas VHDL is used for almost all implementation, Guccione and Keller used JBits to allow faster compile times and
a true run-time partial reconfigurability [185].

The first ASIC acceleration of a *generic SW* algorithm was the Bisp by
Chow in 1991 [80], followed by Samba [187], Kestrel [183], and Swasad [195].
FPGA implementations were proposed by Yamaguchi, Maruyama and Konagaya [431] and by Dydel [132]. Weaver implemented SW matching enhanced
by Leiserson's retiming algorithm after the placement in the FPGA to increase
the clock frequency [412]. Only a few implementations allow affine gaps, such
as those of Oliver and Schmidt [313]. In 2004, Van Court and Herbordt proposed a very versatile implementation with interchangeable components allowing to choice of the type of alignment (local / global), the scoring scheme,
and the usage of a backtracking procedure [96].

8.4 Seed-Based Heuristics

The previous DP equations probably provide the best way to find similarities between two strings. However, their major drawback comes from their
quadratic computation time, preventing them from being used to compare
very large strings such as complete genomes. Efficient heuristics for similarity searches were proposed fifteen years ago by Pearson (FASTA, [324]) and
Altschul (BLAST, [17]). Although today there are numerous other heuristic
algorithms that outperform it, the BLAST software remains the reference
(Table 8.1) algorithm.

In this section, we present these heuristics and how they have been recently implemented in FPGAs. Such approaches combine the advantages of

using efficient algorithms with highly parallel reconfigurable computing implementations.

	Nucleic query (Σ_4)	Proteic query (Σ_{20})
Nucleic bank (Σ_4)	*blastn* / *tblastx*	*blastn*
Proteic bank (Σ_{20})	*blastx*	*blastp*

Table 8.1. BLAST flavors. The program *blastn* directly compares two nucleic strings, whereas the *tblastx* compares the proteic translations on all the 6 reading frames.

8.4.1 Filtering, Heuristics, and Quality Values

When scanning large banks, a DP computation can be viewed as a filtering process that returns the positions of the best alignments (Figure 8.10): objects (sub-sequences) are selected from a large set (complete databank).

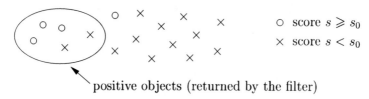

positive objects (returned by the filter)

Fig. 8.10. Schematic view of an heuristic process which returns some objects (positives) among a larger set. Here the heuristic erroneously accepted two objects (false positives) and badly rejected one object (false negative).

The objects returned are $T^{\oplus}$ true positives (objects that really match) and $F^{\oplus}$ false positives (due to the heuristic). Similarly, the non-returned objects are $T^{\ominus}$ true negatives and $F^{\ominus}$ false negatives. The selectivity $S_\ell = (T^{\oplus} + F^{\oplus})/(T^{\oplus}+F^{\oplus}+T^{\ominus}+F^{\ominus})$ measures the raw filtering rate. More interesting are the two quality measures $S_n = T^{\oplus}/(T^{\oplus} + F^{\ominus})$ and $S_p = T^{\oplus}/(T^{\oplus} + F^{\oplus})$. The sensibility S_n is the fraction of the positive results that have been returned. The specificity S_p is the fraction of the interesting results inside the returned results.

8.4.2 BLAST : a 3-Stages Heuristic

Seed-based heuristics massively accelerate the DP computations by omitting certain parts of the matrix calculation. Their basic assumption is that, most of the time, significant alignments keep small words, or *seeds*, conserved in a exact way (as GA in Figure 8.4b). The full DP calculations are run only on neighborhoods of those seeds. BLAST-like algorithms proceed in 3 stages (Figure 8.11):

- Stage 1 looks for exact seeds that appear in both strings. By default, nucleic seeds of size $w = 11$ are searched. Such seeds represent a diagonal in the DP matrix.
- Stage 2 tries to extend each seed with a limited number of substitutions. No insertions nor deletions are allowed, so this extension is along the same DP diagonal. Only extended seeds whose score is greater than a threshold are retained.
- In Stage 3, full DP computations can be done on the extended seeds.

In such a heuristic, the smaller the seeds, the higher the number of hits, and the higher the sensibility. However, using small seeds risks excessive computation in Stage 2. Other methods to generate seeds such as spaced seeds or multiple seeds improve the sensibility even for same w's [273, 312, 384].

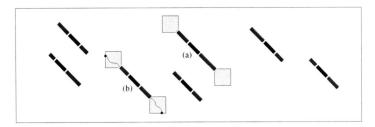

Fig. 8.11. Schematic view of the BLAST 3-stages algorithm. Stage 1: localizes exact seeds (black). Stage 2: extends of the seeds with a few error tolerance (dark gray). The majority of detected seeds doesn't extend at this stage. Stage 3: performs full DP calculations (light gray) on extremely few sequences. Here only the seed (b) leads to a positive sequence.

8.4.3 Seed Indexing

Stage 1 is the most computationally intensive, taking from 70% to 90% of the computing time [245, 302]. Stage 1 computes a *membership problem* [65]: given a seed in the query, does it appear in the bank?

The first solution is to store all the n seeds of one string in an array (with 4^w entries in the case of a nucleic string). The usual information stored in each entry is a pointer to the position in the string, or possibly even several sequence texts, leading to very large array size. [67] suggests methods to pack the array to reduce storage costs.

The second solution is to accept in the membership test a few false positives that will be hopefully discarded in later stages. Seeds are stored in a hash table of size N (Figure 8.12). The probability of having a false positive is then $P(F^{\oplus}) = 1 - (1 - 1/N)^n$. Hash table efficiency can be improved with Bloom filters [53], as used by Krishnamurthy [245]. The idea is to replace a query in a hash table of size M with d queries in a hash table of size $N \leq M$, each query having its own hash function. The object is found if every query succeeded. Now the false positive probability becomes $P(F^{\oplus}) = (1 - (1 - d/N)^n)^d$.

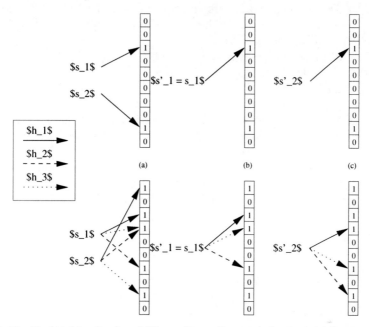

Fig. 8.12. Hash tables (top) and Bloom filters (bottom). In the initialization phase (a), hashing values corresponding to seeds s_1 and s_2 are written to the table. In the reading phase (b), a seed $s'_1 = s_1$ is successfully recognized. The reading phase can have some false positives (c), with the recognition of s'_2 that was not in the original seeds.

A Bloom filter requires the computation of d different hash functions and d different memory accesses. On an FPGA, the computation of hash functions is efficiently parallelized, and simple XOR-based functions are sufficient [337]. Moreover, the hash table can be duplicated in order to parallelize the accesses into it. Dharmapurikar, Attig, and Lockwood were the first to show how to implement Bloom filters on an FPGA in a packet filtering application [114]. The on-board memory on the FPGA (such as Xilinx Block RAM) can be configured to allow multiple concurrent accesses in the same clock cycle. With dual-port memories, only $d/2$ replications of the hashing function is needed to have d concurrent accesses.

8.4.4 FPGA Implementations

As BLAST is very popular among biologists, many attempts have been made to accelerate it, sometimes without asking the biological significance of the algorithm. Surprisingly, many researchers do not question the 3-stage decomposition that was originally designed to accelerate the software computation. Non-FPGA accelerators with clusters of PCs have been developed such as mpiBLAST, an open-source implementation over MPI [104]. The first ASIC implementation of a seed-based heuristic was done by Sigh et al. in 1993 with the BioSCAN architecture [147].

Several FPGA implementations have been independently developed since 2003. At Berkeley, Chang implemented Stages 1 and 2 on the BEE2 system [67], as did Krishnamurthy et al. of Washington University on the Mercury System [245]. At the University of Kansas, Muriki implemented Stage 1 with RCBLAST [302]. The strength of these platforms is the interaction between software and hardware parts, and well-designed pipeline networks for data flow.

Other projects do not exactly follow the BLAST stages, including RDisk, in Rennes (France) which uses a particular seed-based heuristic similar to BLAST's stage 2 [190, 252]. Gardner-Stephen and Knowles (Flinders University, Australia) developed both a new algorithm DASH with better sensibility than BLAST as well as an FPGA implementation to execute it [165, 241].

8.5 Profiles, HMMs and Language Models

When a common family of genomic strings has been identified by pairwise comparisons, real experiments, or human expertise, one can build a *model* representing it. A model can represent a precise domain of several amino acids associated to a function of the protein or a more general profile inferred over all the sequences. Other models can be represented as languages that implicitly describe some structures of the sequence. The models can *generate* sequences, often with a given score or probability that will measure how well the model matches the sequence.

8.5.1 Position-Dependent Profiles

Given a family of similar sequences, one can build a *multiple alignment* as in Figure 8.13 [390]. As far as we know, such algorithms have not yet been implemented on FPGAs, probably because the computations involved are more sequential in nature than in the problems studied in Section 8.3.

Given such a multiple alignment, one establish a *consensus pattern* with the most frequently occurring amino acid at each position. When there are several amino acids with the same number of occurrences, choices, e.g., [PQ], can be used (Figure 8.14a). Many pattern have been designed using such a

```
FOS_RAT      LVQPTLVSSVAPSQ-------TRAPHPYGLP
FOS_MOUSE    LVQPTLVSSVAPSQ-------TRAPHPYGLP
FOS_CHICK    LVQPTLISSVAPSQ-------NRG-HPYGVP
FOSB_MOUSE   LVQPTLISSMAQSQGQPLASQPPAVDPYDMP
FOSB_HUMAN   LVQPTLISSMAQSQGQPLASQPPVVDPYDMP
             ******:**:* **... ::.    .**.:*
```

Fig. 8.13. ClustalW multiple alignment of five proteins [390]. On the bottom line, * denotes a match, and : and . are weaker matches.

syntax, for example, PROSITE, whose syntax allows flexible gaps such as x(3,5) [220].

```
        Y      L      V      P      S      H
        Y      L      A      P      S      H
        Y      L      A      Q      S      H
        Y      A      A      Q      S      H
        Y      M      A      A      S      H
       =========================================
   (a) Y      L      A      [PQ]   S      H
       =========================================
   (b) Y:1    L:0.6  V:0.2  P:0.4  S:1    H:1
              A:0.2  A:0.8  Q:0.4
              M:0.2         A:0.2
```

Fig. 8.14. Consensus (a) and profile (b) from a multiple alignment. The consensus pattern YLA[PQ]SH can be an ancestor of the family. The profile (b) assigns the probability 0.192 to YLAPSH and 0.016 to YMVPSH : the first string is more likely to belong to the family.

More generally, one can keep more statistical information with a *profile* that records the amino acid distribution at each position (Figure 8.14b). How adequately a string matches the profile is computed by multiplying probabilities at each position. Real implementations use log values. Such profiles can easily be encoded in reconfigurable logic with a score or a probability flowing through an array of cells.

8.5.2 Hidden Markov Models

A process is Markovian if the probability distribution of the next states depends only on the given current state (i.e. independent from past states). An Hidden Markov Model (HMM) is a Markov process in which only the output can be observed. A HMM is built from a family of sequences supposed to be part of a Markov process whose parameters (transition probabilities) are unknown, but with a *fixed topology*. Usual topologies include match states, delete

states, and insertion states (figure 8.15). The knowledge of some outputs of
the HMM will reveal the parameters.

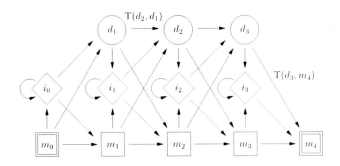

Fig. 8.15. Example of HMM topology, as initially proposed by Haussler [202].

Once the model is trained, one can assign to every sequence a score relative
to the HMM. The Viterbi algorithm reveals the best path through the HMM
and thus the alignment with the consensus.

HMM have been extensively used for twenty years in speech recognition.
The first use of HMM to parse genomic sequences was done by Haussler in
1993 [202] at the University of California.

Hughey proposed in 1993 a parallel implementation of HMM using an
array of processors [218]. In 1999, Mosanya and Sanchez proposed a FPGA
implementation of a similar model called *generalized profile* using a systolic
array. They showed that such an implementation can benefit from on-line
arithmetic [298]. Gupta proposed another FPGA architecture in 2004 [189].

8.5.3 Language Models

The PROSITE syntax described above can express only a subset of regular expressions. One can go further in the Chomsky language hierarchy [78]
while considering the model as a *language* that can be represented in some
way. Searls showed indeed that some features of more complex languages can
benefit DNA parsing [359, 360].

Full regular expressions can be implemented efficiently as finite automata.
The main advantage of the FPGA is that *non-deterministic automata* (NFA)
can map almost directly to hardware using a linear encoding in which one state
maps to one cell. NFA derived from regular expressions were implemented by
Sidhu and Prasanna on a reconfigurable platform [369].

One extension of the usual NFA is the *weighted finite automaton* (WFA),
in which a weight propagates through all the states [296]. The weight can
be arbitrarily chosen and can be not probabilistic. The WFA represent non-
regular languages (Figure 8.16) and are easily implemented in FPGA [167].

In fact, HMM as position-dependent profiles discussed above can be emulated by WFA.

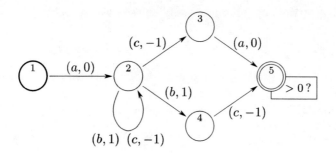

Fig. 8.16. A WFA recognizing the non-regular language $\{w \in \mathcal{L}_1 \mid |w|_b > |w|_c\}$, where $\mathcal{L}_1 = a\,(b|c)^*\,(ca \mid bc)$.

Other non-regular features are *repeats* or *palindromes*, that can be efficiently parsed with systolic arrays [246]. Conti proposed an implementation on FPGA that tolerates mismatch errors [93].

For *generic context-free* parsing, it would be possible to use pushdown automata [214]. A systematic FPGA implementation of a context-free grammar was proposed by Ciressan [82] in a natural language processing application, but currently FPGAs have not been used for context-free parsing in bioinformatics purposes.

As the recognition of context-free grammars is quadratic and can involve non-bounded strings, it is not clear that a generic parser will be especially effective in reconfigurable hardware. However, future reconfigurable implementations could parse *mixed patterns* based primarily on simple patterns or automata, with specific parts that could include computed stochastic parts as HMMs or context-free features such as palindromes or repeats.

8.6 Bioinformatics FPGA Accelerators

Over the last decade, a few dedicated machines have been proposed by the scientific community to speed up genomic computations, especially DNA or protein string comparison. Two categories can be distinguished: VLSI and FPGA approaches. The first category mainly refers to academic prototypes such as, for example, the Bioscan [147], the Bisp [80], the Kestrel [183, 219] or the Samba [187] projects. The second category includes FPGA machines not necessarily devoted to biocomputing but well-suited to support string matching algorithms.

We first give three examples of biosequences algorithm implementation on *general purpose* reconfigurable computing architectures (Splash, Perle and

GenStorm), then we describe in more detail an FPGA-based machine (RDisk) specifically designed for scanning genomic databases. We end by citing two commercial products (BioXL/H and DeCypher) which use FPGA technology and efficiently support the principal bioinformatics algorithms.

8.6.1 Splash

In [210], Hoang presents two systolic implementations of a global distance calculation between two DNA or protein sequences on the Splash 1 and Splash 2 machines [27]. They implemented systolic arrays discussed in the section 8.3 with modulo encoding.

For DNA sequences, 384 systolic cells fit onto a single SPLASH-2 board. Performance of a Splash-2 board, compared to a comparable workstation (SPARC-1), were 1000X higher.

8.6.2 Perle

In [188] the idea is to compute the first stage of the FASTA program [324] on the Perle-1 [407] board. The FASTA heuristic is similar to BLAST and starts by detecting seeds before combining them to construct an alignment. The two first stages were implemented on the board.

A systolic array of 256 cells was implemented on the PERLE-1 board and used for the hit detection step of FASTA. Depending on the length of the query sequence, speedups ranging from 50 to 400 were observed, compared to a SPARC-10 workstation.

8.6.3 GenStorm

The GenStorm project (EPFL, Lausanne, Switzerland) is a dedicated accelerator for biological sequence processing [298]. The original contribution of this effort was to investigate on-line arithmetic and redundant data representation for implementing biocomputing processing. From an architectural point of view, the structure of the GenStorm board does not differ much from the Splash or Perle boards: 9 FPGA components plus local memory are tightly interconnected together, one FPGA being dedicated to interface the board through a VME bus.

The main application tested on the GenStorm board was a generalized profile search algorithm, as described in section 8.5.

8.6.4 RDisk

The RDisk project [252] mainly focuses on scanning large genomic banks. The aim is to scan databases as fast as possible, either for content-based search, pattern matching or profile search. Speedup is obtained by directly connecting

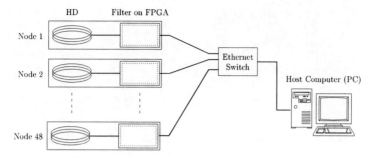

Fig. 8.17. The RDisk architecture. 48 processing boards are interconnected by an Ethernet network. A host computer send the queries and collects the results.

a filtering process near the hard disk data storage, so that filtered results can be transferred quickly to a front-end host computer.

As database scans with dedicated hardware are mainly limited by the time for accessing the data, the RDisk system parallelizes accesses by dispatching the data among a cluster of reconfigurable disks, each of them being able to access the data independently. Obviously, this requires that the genomic database has been loaded onto the distributed storage system, and that each node is able to process data independently. Fortunately, scanning applications are highly parallelizable: a request can be broadcast and processed in parallel across all the disks. When each node has finished, partial results are sent to a front-end processor for further processing if needed.

The complete system is a parallel machine of 48 nodes interconnected by an Ethernet network, as depicted in Figure 8.17. A node is essentially composed of a hard disk and a low cost FPGA (Xilinx Spartan 2). For a cost estimated to be 10 times smaller than a cluster of PCs, performance is ten times higher for complex query searches.

The reconfigurable parallel disk system can be requested to perform various types of searches. It is reconfigured to use filters specifically tuned to each different search. The primary goal of the FPGA component is to efficiently implement the various filters. However, the FPGA is also responsible for managing Ethernet communication and the IDE disk interface. These tasks are independent of the filtering process, and common to all search database applications.

The Spartan-II is configured as a "Reconfigurable System on Chip" (R-SoC). Figure 8.18 shows the components: one part of the hardware resources is devoted to common services – and thus identical for each application – while the second part is customized to each filter request. The common services component uses a (soft) embedded 16-bit microprocessor, connected to an Ethernet, an IDE, and the filter interface. The common services component has been designed to occupy minimum area (1/3 of the Spartan-II) in order to keep the major resources for the filters.

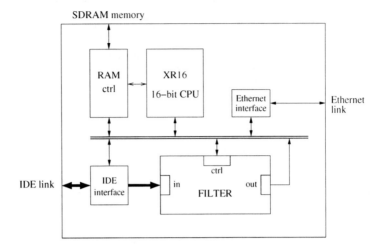

Fig. 8.18. Reconfigurable SoC of the RDisk node. The central component is a 16-bit RISC processor especially designed for the Xilinx family. It is connected to an external 32 MB SDRAM memory through a memory controller. In addition, a small ROM memory, initialized through the bitstream, contains a boot program able to download a larger code from the disk.

The filter is the application specific part of the R-SoC and is easily integrated or interchanged. Each filter implementation must conform to a predefined interface. The internal logic of each filter depends of the algorithm being implemented and its data structures. The embedded RISC processor *sees* only three ports: an input port directly linked to the IDE channel, an output port connected to the system bus, and a control port. To achieve a high throughput out from the disk, the filter gets its input data stream directly from the IDE controller.

Programming the RDisk board for a new application requires the application designer to write (1) a C program for the embedded RISC processor and (2) a VHDL specification of a filter. The C program is the master process and controls all the transactions. Typically, it manages the Ethernet communication, controls the disk transfers, drives the filter, collects the results (data transfer from the filter output to the memory), etc. A C-library of high level primitives allows the programmer to efficiently steer the different elements of the R-SoC system.

The VHDL specification describes the hardware structure of the filter. The filter architectures must take into account the interface constraints: the input port is dedicated to access data from the disk, while the output port is expected to be used only for transferring a small amount of data to the main memory.

8.6.5 BioXL/H

BioXL/H is a high-end hardware accelerator marketed by the Biocceleration Ltd. [441]. It follows previous versions (Biocceleration) designed by Compugen since 1993. BioXL/H accelerates search applications, such as Smith-Waterman, ProfileSearch, FrameSearch and HMM .

A BioXL/H accelerator can house from 4 to 32 processing boards. Each of them contains eight Xilinx FPGA components connected in a ring and 128 MB of memory. All the processing boards are connected through a PCI bus. Also, attached to the PCI bus is an internal hard disk and a fast Internet interface.

Databases are transferred to the BioXL/H accelerator through the network interface. Databases are stored in the board memories and on the internal disk. The search calculations are performed concurrently with the database transfer into board memory and to the internal disk (the "streaming" mode). Databases that can fit into the combined global memory of all boards remain in the memory for subsequent searches (the "memory" mode). Larger sequence databases, such as GenBank, are read from the internal disk in subsequent searches (the "disk" mode).

8.6.6 DeCypher

This FPGA accelerator, dedicated to biocomputing, is designed by the Time-Logic Company, who do not release architectural details of the DeCypher FPGA boards. In general, DeCypher boards include both FPGA components and memory, and can be plugged into standard PC stations through a PCI bus.

Together with the hardware, new software algorithms have been tailored to the embedded FPGA technology. Accelerated versions of BLAST or Smith and Waterman algorithm have been developed, as well as HMM profile analysis. Reported performance on intensive computing genomic applications such as bank to bank comparison are impressive compared, for example, to a 32 CPU Linux cluster.

8.7 Summary

The rapid growth of biotechnologies, especially in large sequencing projects, has lead to an explosion of genomic data. For molecular biologists, this mass of data is potentially a rich source of knowledge with appropriate *in-silico* processes. From a computational point of view, the data being analyzed are primarily text sequences (protein or DNA), and the main task consists of computing similarities. This falls into the string processing computation family which has been studied for a long time, and for which numerous parallel architectures have been proposed. FPGAs are well suited for implementing these regular structures and dedicated reconfigurable accelerators have been designed to speed-up key bioinformatic algorithms. Today, commercial products using FPGAs, derived from previous academic research, exhibit impressive performance compared to parallel microprocessor-based machines.

In this chapter, we have first introduced some important genomic applications to highlight the computer power needed. Then, we focused on the algorithmic problems and their associated hardware implementation:

- the *historical* dynamic programming algorithm has been detailed together with its *systolic* structure;
- more recent works on seed-based heuristics to rapidly search genomic banks have been discussed;
- HMMs and language models to perform more complex computations have been presented. Finally, we concluded with some realizations of FPGA accelerators dedicated to genomic computations.

9

Supercomputing Applications

Acknowledgments: The material in this chapter is derived from a paper "Accelerating Monte Carlo Radiative Heat Transfer Simulation on a Reconfigurable Computer: An Evaluation" by Maya Gokhale, Janette Frigo, Christine Ahrens, Justin L. Tripp and Ronald G. Minnich, published in FPL 2004; and from material in a paper "Acceleration of Traffic Simulation on Reconfigurable Hardware" by Justin L. Tripp, Henning S. Mortveit, Matthew S. Nassr, Anders A. Hansson, and Maya Gokhale, presented at MAPLD 2004.

9.1 Introduction

In contrast to the application classes presented earlier, supercomputing applications are characterized by compute-intensive floating point computations over very large data sets with irregular access patterns. Traditional supercomputing systems contain high performance floating point units, vector pipelines, and/or large numbers of clustered high end workstations.

FPGA architectures with multi-million gates, embedded memory blocks, and hardware multipliers open the door to high performance floating point computation on these devices. Recently, new system architectures have emerged (described in Chapter 3) that use such FPGAs in clustered supercomputers. Applications for these reconfigurable supercomputers are still emerging (as of mid-2005). Early efforts include n-body simulations ([263], [193] and molecular dynamics simulations ([31], [97]). These approaches use reduced precision floating point formats customized to the specific application in order to obtain speedup on FPGAs.

In this chapter we will describe two representative applications that have been mapped to reconfigurable supercomputers. The first application simulates radiative heat transfer in a many-sided 2D chamber. This is a traditional supercomputing application with a large number of floating point calculations. However, as discussed in Section 9.2, the data access patterns are fairly regular, making it possible to partition computation between software and hardware at the loop level of granularity. In this application, single precision floating point is used.

The second application is a large scale simulation of urban road traffic. The application is based on TRANSIMS [2], which is used by the Department of Transportation for planning and analysis of urban road traffic patterns. The application divides the road network into millions of road cells, forming a cellular automaton. Small fixed point integers are used to encode road cell state. The scale of the data sets – Houston, Portland, OH, and Chicago have been simulated – make this a supercomputing scale computation.

9.2 Monte Carlo Simulation of Radiative Heat Transfer

As demonstrated in previous chapters, reconfigurable computing with FP-GAs has shown speed-ups of one to two orders of magnitude on data- and compute-intensive processing tasks involving fixed point computation on small integers. Floating point computation was not mapped to FPGAs due to the large operand size (32- or 64-bit) and excessive area consumed by floating point arithmetic units on configurable logic cells. Earlier work [311] found that FPGAs were not fast enough to be competitive with general purpose processors for floating point. Recently, that limitation of FPGAs appears to be receding: 3–10 million gate FPGAs with embedded processors, memories, and arithmetic units have become available, making it feasible to consider a broader range of applications than before, including those requiring floating point operations [346]. Studies comparing floating point performance of FPGAs vs. high performance microprocessors [399] suggest that peak FPGA floating-point performance is growing significantly faster than peak floating-point performance for a CPU. Other studies [77,363] also suggest that modern FPGAs may be competitive with microprocessors on matrix operations such as matrix multiply and LU decomposition.

FPGAs offer several advantages when used to calculate floating-point operations. First, FPGAs offer a high degree of flexibility, where they can provide a customized solution for a given floating-point algorithm. Second, due to the available concurrency, an FPGA can provide a floating-point solution that is faster than a general purpose processor. Third, FPGAs are based on SRAM, and thus they track the more aggressive semiconductor fabrication improvements for memory rather than microprocessors.

Offsetting those advantages are the slow clock speed relative to microprocessors and the relatively large area required by floating point operands and operations, which limits spatial parallelism opportunities. In addition, it is well-known in the supercomputing community that measures of peak performance and dense matrix kernel operations are far from accurate predictors of realized performance of a complete application. Memory access patterns, control flow, and inter-processor communication result in actual performance that is well below peak. For example, applications run on a cluster supercomputer often realize no more than 50–80% of theoretical peak [393], reducing a 30 TFLOP machine to 15 TFLOPs.

In this experiment, the performance of FPGA-based floating point computation on a real application is quantified by mapping the compute-intensive inner loop onto the Xilinx Virtex-II and Virtex-II Pro family FPGAs and comparing performance to comparable microprocessors.

9.2.1 Algorithm Description

This algorithm models the geometry of a laser isotope separation (LIS) unit to accurately determine the radiant exchange factors among the surfaces. This is an important component of the isotope separation process simulation.

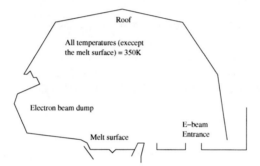

Fig. 9.1. Test Geometry for Radiative Heat Transfer

The simulation is a Monte Carlo application that traces a large number of photons emitted from the surfaces of a 2-D enclosure (Figure 9.1). The simulation records how many photons emitted from each surface i were absorbed at surface j. This information is used to compute a heat transfer coefficient between each pair of surfaces, i and j. It is a Monte Carlo application because it uses random values to determine characteristics of an emitted photon's path and because it traces a large number of photons.

In the algorithm, **N** photons are emitted (with randomly chosen characteristics) from each surface of an **m**-sided polygon. The algorithm follows the path of each emitted photon. It identifies the surface of intersection, which is the most computationally intensive portion of the algorithm. Next, a random number determines whether the photon is absorbed into the surface, reflected off of it, or transmitted through it. The photon is followed until it is transmitted, absorbed or lost. This algorithm is designed to calculate intersections assuming a convex chamber.

A parallel version of the algorithm distributes work at the "task" level. The pseudo-code for each task is summarized in Figure 9.2. In loop "a", a task iterates through the **m** surfaces of the polygon and traces the **N** photons

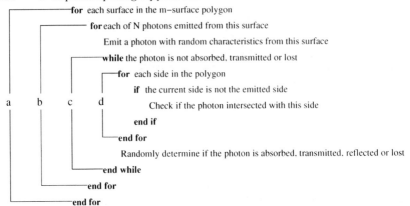

Fig. 9.2. Radiative Heat Transfer algorithm loop structure. Loop "d" is implemented on the FPGA.

emitted from each surface. For each surface, a **for** loop ("b") iterates through each photon emitted, then an inner **while** loop ("c") checks if the photon is still active before following it to its next surface intersection. Inside the **while** loop, the innermost **for** loop ("d") computes the surface intersection, then the random number generator determines if the photon is absorbed, reflected, transmitted or lost.

There are several different ways that this code could be parallelized. However, most of the parallel mappings are not suitable for FPGA implementation. Parallelism at the task or surface level is too coarse, and does not fit on currently available FPGAs. At the **while** loop level, tracing one photon's path until it is not active may be possible in terms of fitting on an FPGA, but there are dependencies carried between loop iterations that make the implementation more complex and limit parallelism. At the inner **for** loop level, where the algorithm checks for the surface of intersection, the code is straightforward to realize on an FPGA, since the loop iterations are independent of each other and can be spatially replicated on the FPGA.

In addition, this inner **for** loop is the most computationally intensive portion of the program. With **N**=5000 and **m**=37, a Pentium IV Xeon 3-GHz workstation spends 86% of the algorithm time executing the inner **for** loop. The C code inside this loop is included in Figure 9.3. All the variables used in the arithmetic computations are floating-point.

The original program was written for double precision floating-point. However, there is not a significant difference in the scientific results from the algorithm when using single versus double precision. The number of photons absorbed differed by only .0025% in the single precision version as compared to the double precision version. Experimental results were obtained for 64, 32, and 16-bit floating point implementations of the loop.

Several commercial [304, 334] and open source [38, 346] libraries are available for creating floating-point circuits. For our implementation of the radiative heat transfer algorithm, we chose the FPLibrary, a VHDL library of

```
float x1[NSM], x2[NSM], y1[NSM], y2[NSM], delx[NSM], dely[NSM], sqln[NSM], rhs[NSM];

delxs = delx[s];  delys = dely[s];  rhss = rhs[s];
x1s = x1[s];  y1s = y1[s];  x2s = x2[s];  y2s = y2[s];  sqlns = sqln[s];

/* compute intersection points*/
det = ex*delys - ey*delxs;
absdt = fabs(det);
if(absdt <= epsdet0) det = epsdet0;
dtinv = 1.0/det;
xi = dtinv * (delxi*rhse - ex*rhss);
yi = dtinv * (delyi*rhse - ey*rhss);

/* test for intersection between surface endpoints*/
ssq  = (xi - x1s)*(xi - x1s) + (xi - x2s)*(xi - x2s)
     + (yi - y1s)*(yi - y1s) + (yi - y2s)*(yi - y2s);
if(ssq <= sqlns) {
  intersect_side[s] = 1;  /* s is the intersected side */
  else intersect_side[s] = 0; /* break here in the software version */
}
```

Fig. 9.3. Radiative Heat Transfer code implemented on the FPGA

hardware operators for floating-point computation, developed by the Arénaire project [113]. The FPLibrary meets three important qualifications. First, it is written in VHDL in a platform-independent manner. This allows designs to be easily targeted to different FPGA architectures. Second, the library implements add, multiply and divide floating point operations which are required for this algorithm. Third, the modules and floating-point types have parameterizable bitwidths, so that we can easily program the library for single, double or arbitrary sized floating point types. FPLibrary is used to leverage the advantages of FPGAs to implement the core of a supercomputing application.

9.2.2 Hardware Implementation

The hardware implementation has been mapped to the Virtex-II and Virtex-II Pro FPGAs. An initial loop pipeline was generated from the Streams-C compiler [171] on an integer version of the code. The generated pipeline was then converted to use floating point modules, and manually optimized to maximize pipelining.

As seen in Figure 9.3, each iteration of this loop performs calculations relative to one of the surfaces of the convex shape. Some variables are invariant across loop iterations (e.g., **epsdet0**) while others assume unique values for each loop iteration, as shown by the array index **s**, for example, **delxs**, **delys**, and **rhss**. The latter variables are assumed to reside in Block RAMs.

Figure 9.4 shows the pipelined hardware implementation of the loop. The design is an 11 stage pipeline utilizing the floating point operators from FPLibrary [113]. It consists of 12 multiply, 3 add, 7 subtract, 1 divide and 2 compare modules. Each of the floating point operators is also pipelined. The multiply has a latency of 4 cycles, the add/subtract has three 3 cycles of latency, division has 15 cycles, and comparison has 1 cycle. The total latency of the

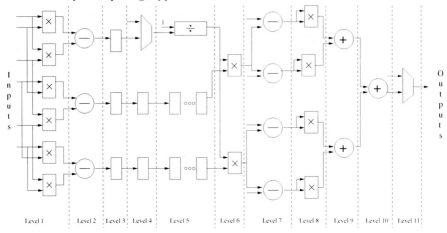

Fig. 9.4. FPGA Implementation

11 stage pipeline is 41 cycles. There are 2 intermediate registers that need pipelining from Level 4 through Level 5. This data synchronization requirement introduces 32 additional 34-bit registers into the design.[1] For clarity, only two registers are shown in in the Figure 9.4 in Level 5, but there are 15 registers for each operand, for a total of thirty 34-bit registers at Level 5.

For this implementation there are eleven inputs to the pipeline – six inputs are consumed in Level 1, four at Level 7 and one at Level 10. The data is stored in two 204-bit by 512 deep, dual-port Block RAMs. Memory reads are scheduled so that values arrive at Level 7 and at Level 10 at exactly the cycle they are consumed. By scheduling the reads in this way, we avoid the overhead of fully pipelining the 5 inputs that are needed at Level 7 and Level 10. The latter approach introduces an extra 27 cycles × 4 registers (Level 7) plus 40 cycles × 1 register (Level 10), or 112 + 40 = 152 34-bit registers into the design. If these 152 registers were included in the design, there would be a 1% increase in area used on the Virtex-II.

9.2.3 Performance

The performance of the inner loop hardware implementation is compared to several Pentium IV Xeon (P4) systems. The hardware circuit was mapped to the Virtex-II, speed grade 4 and Virtex-II Pro, speed grade 6.

Workstation Performance

For performance comparisons with the FPGA the innermost loop of the application is timed, This loop is the iteration over **m** surfaces for a single photon, searching for an intersection. The static instruction count of this loop count

---

[1] The FPLibrary adds a 2-bit tag to each floating point register.

is 130 instructions: 61 floating point instructions, 9 branches, 73 instructions which reference the stack (including floating load/store to stack for locals), and only one integer instruction (the loop counter). All the instructions and data for this loop fit into the Level 1 cache (the fastest cache level), and hence could be expected to run at maximum speed on the CPU.

Timing measurements of the inner loop are easily perturbed due to the small instruction count of 130 instructions. Obtaining an accurate measure of this loop represents a challenge, since traditional profiling tools such as gprof are only acceptable for function-level timing, and do not provide an extremely accurate measure of the inner loop. However, on the Pentium and later processors there is a timer register, called the Time Stamp Counter (TSC), which measures processor ticks at the processor clock rate. This 64-bit read-only counter is extremely accurate, as it is implemented as a Model-Specific Register inside the CPU. The overhead of using this register is very low. On a 1.7-GHz P4 the TSC runs at 1.68 GHz and has a resolution of 595 picoseconds; on a 3-GHz P4 the TSC has a resolution of 333 picoseconds.

The TSC counter was used to measure the inner loop of the application. C code was added using the gcc asm statement, which produces in-line assembly code to read the TSC at the start and end of the loop code. We performed measurements both in the application itself, and by extracting the inner loop and running it many times. As expected for this loop, the performance varied with the CPU being used, with the fastest CPU being the 3-GHz P4.

We tested both the Intel compiler v7.0 and gcc v3.2. The gcc compiler provided the best performance results with –O3 optimization level. Timing for one iteration of the inner loop, shown in Figure 9.5, ranges from 60 ns to 104 ns. It is important to note that the time is an average, since in the sequential version of the loop body, there is opportunity for early exit from the loop.

FPGA Performance

Synplicity's Synplify was used to synthesize the inner loop to Xilinx FPGAs. Placement and area results were obtained using Xilinx ISE 6.1. The results for one iteration of the inner loop on the Virtex-II and Virtex-II Pro FPGAs are shown in Figure 9.5. On the V2-6k, only 20% of the Look Up Tables (LUT) are used by the loop body. However, all the multipliers are used, and therefore only one instance of the loop body can fit on this part. In contrast, the larger Virtex-II Pro parts can fit three pipelines of the inner loop, resulting in a higher degree of spatial parallelism. The speed up row calculates speed up relative to a 3-GHz P4. The hardware calculation assumes a steady state pipeline in which a result is delivered every clock cycle; with three pipelines three results are delivered every clock cycle. These results do not include the time to write the parameters into Block RAM, which can be hidden by double buffering: parameter block $i+1$ can be loaded while the hardware is processing parameter block i.

	V2-6k		V2p100		1.7-GHz P4	2.4-GHz P4	3-GHz P4
# Pipelines	1	1	2	3	-	-	-
Execution Time (ns)	29.9	16.7	7.89	5.78	104	74	60
%Area (LUTs)	20	15	33	50	-	-	-
%Multiplers	100	32	64	97	-	-	-
Latency (cycles)	41	41	41	41	-	-	-
Speed up	2.01	3.59	7.61	**10.37**	0.58	0.81	1

Fig. 9.5. FPGA vs. Workstation performance for Inner Loop. Speed up compared to the P4 3-GHz System.

In terms of technology generation, the V2-6k and 1.7-GHz P4 are comparable. The V2-6k hardware implementation outperforms the 1.7-GHz Pentium by a factor of 3.48. For the more recent generations of FPGA and microprocessor (V2p100 and 3-GHz), the single pipeline speed up is slightly better – 3.59×. In addition, it is possible to fit three pipelines on the V2p100, yielding a speed up of 10.37.

As noted above, the hardware design is highly pipelined. The pipelining allows a relatively high clock frequency for the design, at the cost of high latency – 41 clock cycles before the first result appears. For a large number of surfaces, the effect of pipeline latency diminishes. 150,000 surfaces are desirable for this particular simulation, so the pipeline latency effect is negligible.

Lastly, if we analyze the granularity of the input data width as shown in Figure 9.6, placement results show that for a 16-bit word width, the area utilization across the Virtex Family is 5% to 8% which allows 10 to 20 instantiations of the inner loop to run concurrently on the FPGA. For larger bit widths, fewer parallel versions of the loop can fit onto hardware, for example with 32-bits 3 pipelines fit. As expected, the run-time clock speeds are faster for smaller bit widths. The results show that on the Virtex-II Pro family, 32-bit operations are only slightly more expensive than 16-bit, while 64-bit incur a much higher penalty both in area and clock speed. As the graphs show, the 64-bit version of the application does not fit on the V2-6k.

Discussion

The results show that FPGA hardware outperforms a comparable generation of microprocessor by up to 10.37× on an application-specific single-precision floating point pipeline. There are several points to note.

First, the FPGA implementation must execute all loop iterations of the inner **for** loop. The software timing is an average number: many times the software breaks out of the loop without completing all iterations, as the last **if** statement of Figure 9.3 contains a **break** in the software version of the loop. If all loop iterations were executed, the FPGA speed up would be even greater.

Second, this application fits well in the L1 cache of the microprocessor. A more data-intensive application would better use the strengths of the FPGA:

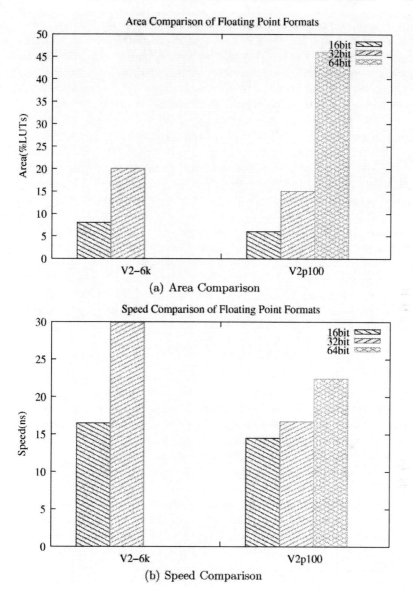

Fig. 9.6. Placement results for a single pipeline 16, 32, and 64-bit implementations of the inner loop.

greater memory bandwidth and better performance on data-intensive computation.

Third, the tractability of an application kernel to pipelining, especially long pipelines, is crucial to achieve performance. The highest performance floating point operators are heavily pipelined, so there is substantial cost in starting up and breaking up the pipeline. Like vector processors, the application-specific pipeline on the FPGA shows the best performance when the algorithm has many iterations with minimal data-dependent branching. In this application, the vector length is very large, and thus the latency is negligible. This application also has the advantage of little data-dependent branching. Although predication can be used to reduce the impact of branching, area costs increase by having both the **then** and **else** bodies instantiated on the chip.

Fourth, the floating point library we used in this experiment is neutral to FPGA technology. In fact, we were able to synthesize it to several different families, including the Altera Stratix. Technology-specific floating point cores such as Quixilica yield smaller area and faster clock rate. On the minus side, other floating point libraries, including Quixilica, have even higher operation latencies. For best performance, embedded hard floating point units in a fabric of reconfigurable logic would, of course, be desirable.

Finally, it is important to compare the performance of the application-specific pipeline, with a mix of different floating point operators and branching constructs, to peak performance results cited by others. While theoretical peak numbers are useful to gauge feasibility, a floating point intensive super-computing application gives us more accurate performance results.

9.3 Urban Road Traffic Simulation

In addition to intensive floating point calculations, another defining feature of supercomputing applications is the need to process large data sets. In the road traffic simulation application we highlight the use of FPGAs on discrete simulation problems that process large data sets.

Modern society relies on a set of complex, inter-related and inter-dependent infrastructures. Los Alamos National Laboratory has over the past ten years developed a sophisticated simulation suite for simulating various infrastructure components, such as road networks (TRANSIMS [2]), communication networks (AdHopNet [30]), and the spread of disease in human populations (EpiSims [149]). These powerful simulation tools can help for example policy-makers understand and analyze these inter-related systems and support decision-making for better planning, monitoring, and proper response to disruptions. TRANSIMS, for example, can simulate the traffic of entire cities, with people traveling in cars on road networks. It is based on interacting cellular automata (CA), and uses large computer clusters for efficient computation.

In this work we study the acceleration of the road network simulation through an FPGA implementation. Since the simulation is parallel, with independent agents that make decisions based on local knowledge, it seems natural to map to the large-scale spatial parallelism offered by FPGAs. The high degree of regularity found in the road network is another fact making this a well suited application. In contrast, other networks such as ad hoc wireless communications are much more irregular and dynamic.

9.3.1 CA Traffic Modeling

The TRANSIMS road network simulator can best be described as a cellular automaton computation on a semi-regular grid or cell network. The simulator consists of two parts,

- a planning/routing module that populates the simulation with cars and decides on the route each car takes, and
- the road network micro-simulator, which actually simulates the movement of the cars.

In this work the focus is on the micro-simulator. In the micro-simulator, the city road network is split into nodes and links. Nodes correspond to locations where there is a change in the road network such as an intersection or a lane merging point. Nodes are connected by links that consist of one or more unidirectional lanes. Each lane is divided into road cells. The TRANSIMS road cell is 7.5 meters long. One cell can hold at most one car. A car travels with velocity $v \in \{0, 1, 2, 3, 4, 5\}$ cells per iteration step. The positions of the cars are updated once every iteration step using a synchronous update. The maximum speed of a car is cell dependent, but it is at most 5. Each iteration step advances global time by one second. The basic driving rules for multi-

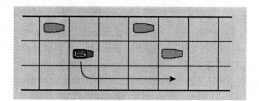

Fig. 9.7. CA traffic in TRANSIMS

lane traffic in TRANSIMS consists of four steps. In each step we consider a single cell i in a given lane and link. Note that the model allows passing on the left and the right. To avoid cars merging into the same lane, cars may only change lane to left on odd time steps and only change lane to the right on even time steps. This convention, along with the four driving rules described below, produces realistic traffic flows as demonstrated by TRANSIMS.

Local driving rules

The four basic driving rules of the micro-simulator are given in the following. We let $\Delta(i)$ and $\delta(i)$ denote the cell gap in front of cell i and behind cell i, respectively.

1. *Lane Change Decision:* Odd time t: If cell i has a car and a left lane change is *desirable* (car can go faster in target lane) and *permissible* (there is space for a safe lane change) flag the car/cell for a left lane change. The case of even numbered time steps is analogous. If the cell is empty nothing is done.

2. *Lane Change:* Odd time t: If there is a car in cell i, and this car is flagged for a left lane change then clear cell i. Otherwise, if there is no car in cell i and if the right neighbor of cell i is flagged for a left lane change then move the car from the neighbor cell to cell i with probability p_α. The case of even times t is analogous.

3. *Velocity Update:* Each cell i that has a car updates the velocity using the two-step sequence:
   - $v := \min(v + 1, v_{\max}(i), \Delta(i))$ (acceleration)
   - If $[\text{UniformRandom}() < p_{\text{break}}]$ and $[v > 0]$ then $v := v - 1$ (stochastic deceleration).

4. *Position Update:* If there is a car in cell i with velocity $v = 0$, do nothing. If cell i has a car with $v > 0$ then clear cell i. Else, if there is a car $\delta(i) + 1$ cells behind cell i and the velocity of this car is $\delta(i) + 1$ then move this car to cell i. The velocity update pass (3) guarantees that there will be no collisions.

All cells in a road network are updated simultaneously. The steps 1–4 are performed for each road cell in the sequence they appear. Each step above is thus a classical cellular automaton Φ_i. The whole combined update pass is a product CA.

$$\Phi = \Phi_4 \circ \Phi_3 \circ \Phi_2 \circ \Phi_1,$$

where product is function composition. Note that the CAs used for the lane change and the velocity update are stochastic CAs. The rationale for having stochastic braking is that it produces more realistic traffic. The fact that lane changes are done with a certain probability avoids slamming behavior where whole rows of cars change lanes in complete synchrony.

9.3.2 Intersections and Global Behavior

The four basic rules handle the case of straight roadways. TRANSIMS uses travel routes to generate realistic traffic from a global point of view. Each traveler or car is assigned a route that it has to follow. Routes mainly affect the dynamics near turn-lanes and before intersections as cars need to get into a lane that will allow them to perform the desired turns.

To incorporate routes the road links need to have IDs assigned to them. Moreover, to keep computations as local as possible, cells need to hold information about the IDs of upcoming left and right turns.

The following describes the extension of the four basic driver rules to handle turn-lanes and intersections.

Modification of the lane change rule:

We consider a car in cell i. As before, lane changes to the left/right are only permissible on odd/even numbered time steps. We refer to the adjacent candidate cell as the target cell.

1. If the link ID of the target cell matches the next leg of the travel route a lane change is desirable (desirable turn-lane).
2. Else, if the target cell has a link ID that does not match the next leg of the route and it differs from the current link ID of the route, a lane change is not desirable (wrong turn).
3. Else, if the current cell's nextLeftLink (nextRightLink) ID matches the next leg of the route and the simulation time is an odd (even) integer, a lane change is desirable (prepare for turn-lane or intersection).
4. Else, apply the basic lane changing rule described above.

Note that this handles lane changing prior to turn-lanes as well as intersections.

Intersection Logic

An intersection has a number of incoming and outgoing links associated to it. A simplified set of turning rules (assuming a four-way intersection) are as follows:

1. Only cars in an incoming left(right)-most lane of link can turn left(right). A car that turns left(right) must initially use the left(right)-most lane of the target link.
2. A car in any incoming lane can go straight. A car that goes straight must use the same lane number in the target link as it used in the incoming link. It is assumed that the lane counts for the relevant links agree.

More intricate intersection geometries can of course occur but the basic idea remains the same. When intersections are close it is natural to modify the first rule: when a left turn is followed by an immediate right turn the rightmost lane is chosen as target lane for the left turn.

An intersection has a set of immediate adjacent road cells. We refer to these as the *intersection road cells.* The intersections operate by dynamically assigning the front and back neighbor cell IDs of the intersection road cells. This allows us to naturally extend the driving rules for multi-lane traffic to intersections without any modifications. The subset of the intersection road cells that come from incoming links have their front neighbor cell set to zero by default. The same holds for the back neighbor of the intersection cells belonging to outgoing links. The intersections operate by establishing front/back pairs between cells to accommodate the routes. Stop intersections and traffic signal intersections impose additional constraints on which cars are allowed to drive at what times by controlling the corresponding connections.

In order to evaluate the benefit of FPGA acceleration, a subset of the road traffic CA consisting of straight lane traffic was implemented in hardware. Two approaches were developed, a constructive method in which each road cell is physically instantiated in hardware, and a streaming method, in which cell state store in memory is streamed through a hardware update engine.

9.3.3 Constructive Approach

The constructive approach to traffic simulation instantiates a separate simulatable road cell for every road cell in the traffic network. The road cell provides its current state to its neighbors so that all the cells in that local neighborhood can calculate their next state. Figure 9.8 outlines the structure of a basic road cell.

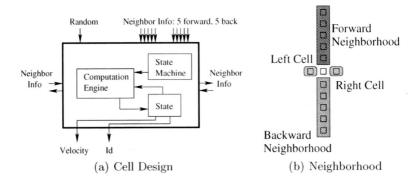

(a) Cell Design   (b) Neighborhood

Fig. 9.8. Road Cell Design and Cell Neighborhoods

The road cell consists of three main parts: the computation engine, the state and a state machine. The state machine drives the computation engine using the current state and inputs from external road cells to compute the road cell's next state. The local driving rules define operations of the computation engine.

The four rules for traffic simulation are executed using six different states in the state machine. Figure 9.9 shows how the different steps in the rules are executed in the state machine. Each rule in the computation engine requires a single cycle to calculate except for *Velocity Update*. The velocity update rule has three separate operations, each taking a cycle, to calculate its two steps.

In the LaneChange state, the computation engine calculates the lane change decision (Rule 1). To do this the $\Delta(i)$ and $\delta(i)$ are calculated from the forward and backward neighborhoods. Likewise the neighbors in the left and right neighborhoods execute the same calculation. The computation engine then determines whether it is permissible for a car to come to this lane and whether the current car desires to change lanes. These results are used in the LaneMove (Rule 2)state to actually perform the lane change. Both lanes have to agree that is is both permissible (where we are going) and desirable (if the gap ahead of us is worse than our neighbors) for a lane change to happen.

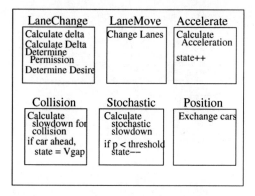

Fig. 9.9. Computation required for the six states

Rule 3 requires three states, Accelerate, Collision, and Stochastic. In the Accelerate state, a car's velocity is calculated using the following formula: $v_{next} = \min(v + 1, v_{max}(i))$. $v_{max}(i)$ is the maximum velocity for this particular road cell, which may be lower than the global v_{max} (e.g., a local speed limit).

The Accelerate state is followed by the Collision state which ensures that the next state does not exceed the gap ahead of the car. It determines $v_{next} = \min(v_{next}, \Delta(i))$. This prevents the car from accelerating into a car in front of it—avoiding a collision.

The final step of the velocity update determines if the car should randomly slow down. This stochastic step provides some realism in the behavior of drivers and makes their speeds less predictable. If a random value is less than a threshold, p_{break}, then its speed will be lowered as described in Section 9.3.1.

After the velocity update rule is finished, the state machine executes an update of the car positions. To do this, a cell determines if a car exists in its backward neighborhood that has a velocity that will bring it to this cell's location. If it does, then the cell sets its velocity and car id to the arriving car. Otherwise, if no car is arriving at this cell, the cell sets its velocity and car id to zero.

The FPGA implementation is written in VHDL and synthesis was performed by Synplify v7.6. The resultant EDIF description was passed into Xilinx ISE v6.2 to produce the results reported.

The results for the constructive approach for multi- and single-lane circular traffic are described in Table 9.1. The hardware implementation of single-lane traffic has only four states, since single-lanes do not require the extra hardware for lane changes. The two-lane implementation that includes the hardware to perform lane changes is 63% larger in area. As the table shows, both Xilinx chips can hold (at least) 650 road cells. The Virtex-II Pro could hold more cells, but with 650 cells can operate at a faster clock speed.

Table 9.1. Constructive Method Design Results

	One-lane		Two-lane	
	V2-6k	V2p100	V2-6k	V2p100
Cells	650	650	400	640
LUTs/Cell	104	97	169	128
Clock(MHz)	48.68	64.17	35.53	62.8
Slices	33790	31576	33790	40973
(% of Slices)	(99%)	(71%)	(99%)	(92%)

Table 9.2 compares the results for the two-lane traffic implementation achieved by the Xilinx XC2V6000 (V2-6k) and the XC2VP100 (V2p100) to a software implementation running on two different Xeon processors. The speedup reported is relative to the 3-GHz Xeon. The V2-6k simulates the road cells at a rate $175.6\times$ the Xeon. This speedup comes primarily from the fact that the FPGA implementation is executing all cells concurrently, and the software implementation, which may have instruction level parallelism, calculates each cell individually.

Table 9.2. Constructive Method Comparison for Two Lanes

	V2-6k	V2p100	2.2-GHz Opteron
Cells	400	640	2 Million
Cells/sec	2.37×10^9	6.70×10^9	5.7×10^6
Speedup	415.8	1175.4	1.0

Despite the large speedup that is possible using the constructive approach, the FPGA can only handle a small number of road cells. Using the data from the Portland TRANSIMS study, we know that there are roughly 6.25 million road cells. Simulating Portland would require at least 12,400 FPGAs to simulate the entire city. The largest multi-FPGA systems provide 10's to 100's of FPGAs. This limitation in scalability using current technologies led to the development of a streaming approach to calculate the traffic simulation.

9.3.4 Streaming Approach

The constructive hardware approach does not scale well because each road cell must be physically instantiated on the FPGA, requiring a very large number of FPGAs to simulate even a moderate size city. An alternative approach is to let a computational unit, an *update engine*, process a *stream* of road data and subsequently output a *stream* of updated data. In this way, the update engine sweeps across the road data, and the number of road cells is no longer limited to available FPGA area. The streaming approach is only limited by the size of the memory to hold road cell state and the associated access time, and the hardware design becomes inherently scalable and can thus handle large-scale road networks.

The streaming approach focuses on straight lane computation in hardware, leaving intersections and merging links to be calculated in software. Most importantly, this strategy means that all road plan decisions are handled by the software, and the hardware processing is governed by a simple, homogeneous set of traffic rules.

In a basic implementation of streaming, data representing the road network in the previous state is loaded into the input SRAM from the host. The data is subsequently fed into the update engine against the flow of traffic, starting from the end of the link. In the case shown in Figure 9.10, there are four lanes of data per link. Each lane has its own input and output SRAM, and each car both changes lanes and moves forward inside the internal logic engine before it is written with its new road cell and new velocity information to its new lane's output SRAM. Since the road cells are processed in order, and one at a time, the address generators are counters (only enabled when valid data has come out of the SRAM on the previous read).

Due to the large size of the data set relative to available on-chip memory, road cell state is stored in memory external to the FPGA. To process the straight road segments of Portland, two nodes of a Cray XD1 are required, each with 16MB of memory on the FPGA board.

At each simulation time step, the hardware road cell processor swaps its input and output memory banks, eliminating the need for data copy. A small amount of data transfer between the FPGA and host is necessary for information exchange to and from the software updating the merge nodes, intersections, and overlap road segments between software and hardware.

Inside the logic engine, the road cells are first scanned for cars. If a car is found, the car must have its position updated based on its attributes from the previous state and the location of the surrounding cars.

The cars in the first v_{max} cell layers of a lane can not be updated (either by changing lanes or by changing velocity) since there is not enough information available. For this reason we define the overlap region as the end v_{max} cell layers of a link. These cell layers are the first segments processed, and while cars can be moved into these road segments, they cannot be moved from them. The software driven intersection/link-merge updater will update the position and velocity of these cars.

Changing lanes and changing velocity are executed as two separate calculations. Position update must be calculated after changing lanes, since it directly depends on the location of the cars. A pipeline calculating the lane

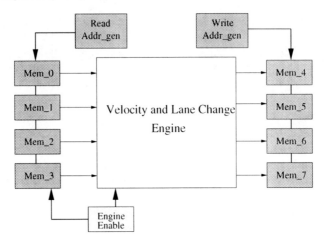

Fig. 9.10. Structure of a straightforward streaming implementation.

changes and velocity updates allows results to be read and written every clock cycle.

1. **Changing Lanes:** The functionality works exactly as described by the four rules above. However, since a car must have complete information about the road segments in front of and behind it to a distance of v_{max}, it is only immediately outside of the overlap region where the cars will have enough information to make a lane change decision. Therefore, for a car to have enough neighbor information to make a lane change decision, the decision must be made v_{max} clock cycles after the car's information has come out of the SRAM. Until then, the car's information moves though a shift register, and its existence is used as neighbor information for the cars ahead.

2. **Changing Position:** To change position, the first step is calculating the new velocity based solely on the car's old velocity, and the existence of any car v_{max} or less in front of the car in its own lane. After the new velocity is calculated, the car does not move to the appropriate road segment immediately. Instead, it moves into a shift register (see Figure 9.11).

   It continues to move one cell per clock cycle through the shift register's lookup blocks (see Figure 9.12) until the point where its newly calculated velocity matches its distance from the change state block. At this point, the change state block imports all car information from the car designated to be moved out of the lookup buffer to the new destination road segment. This allows for the new destination road segment containing the moved car data to be written into the output SRAM on the next clock edge.

The streaming implementation was written in VHDL and placed in VHDL wrapper provided by Cray. The wrapper includes interfaces to the "RapidArray" interconnection network and the external quad data rate memories. The

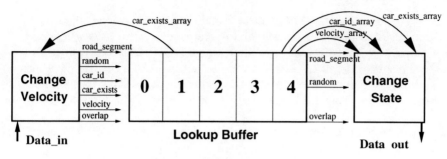

Fig. 9.11. The Position Update

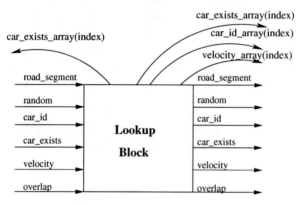

Fig. 9.12. The Lookup Block

streaming design was synthesized using Xilinx XST and the bitstream generated by Xilinx ISE v6.2.

Table 9.3. Comparison of Streaming with Software Simulation

	V2p50	2.2-GHz Opteron
Slices	1857	-
Clock(MHz)	180	2199
Cells/sec	7.2×10^8	5.7×10^6
Speedup	126.3	1.0

The results shown in Table 9.3 were timed using a timer register, called the Time Stamp Counter (TSC), which measures processor ticks at the processor clock rate. The 64-bit read-only counter is extremely accurate, as it is implemented as a Model-Specific Register, inside the CPU. The overhead of using this register is extremely low and the TSC register on the 2.2-GHz Opteron has a resolution of 450 picoseconds.

The design on the FPGA includes four streaming engines (limited by the number of available memories) and operates at a rate 126.3× the speed of a comparable software version running on a 2.2-GHz Opteron. Table 9.4, which includes the cost of transferring data to and from the FPGAs, gives a more accurate speedup of 12.8× faster than software alone. This speedup has been estimated by using 1.3 GB/s as the transfer rate, which Cray has been able to achieve.

Although the streaming implementation is a factor of 100 slower than the direct approach, it is still enough of an improvement to provide significant overall speedup. Additional speedup is still possible with more FPGA boards. The most crucial limiting factor in this implementation is the number of memory banks on each board; additional banks would allow us to increase the number of simultaneous data streams. In fact, with the current design, one compute engine requires less than 2% of chip area. Since each compute node has four concurrent memories, it is advantageous to instantiate four parallel engines, but already at this moderate level of parallelism, we run into a bandwidth bottleneck.

The hardware performs extremely well with the straight lane segments, which make up 70–90% of the road segments in a given simulation. FPGA aided simulation done in the scalable, streaming approach may be the fastest way to do extremely large metro-area traffic simulations, especially in light of the advances being made in combined microprocessor/FPGA computing systems. The cellular nature of the road segments meshes well with hardware, and a combined hardware/software approach for the full-fledged simulation fits each of their computational strengths.

Table 9.4. Comparison of Streaming including Communication Costs

	V2p50	2.2-GHz Opteron
Cells/sec	1.96×10^8	5.7×10^6
Speedup	34.4	1.0

Acceleration of TRANSIMS opens the door to a whole range of simulations where FPGAs or other dedicated hardware can provide computational speedup. Many simulations systems today, such as EpiSims, have similar structure to the one found in TRANSIMS: There are highly complex computations best suited for software and a large collection of structured simple calculations as in the road network simulator. The TRANSIMS accelerator provides a prime example as to how FPGAs can aid a large class of large-scale simulations.

9.4 Summary

While the strengths of reconfigurable computers in signal and image processing applications have been apparent since their inception, reconfigurable
accelerators using multi-million gate-equivalent heterogeneous systems-on-a-
chip have expanded the application space. Supercomputing applications using
floating point data demonstrate significant speed-up over high performance
workstations. Large data sets can also be processed by these accelerators as
interconnection network bandwidth between microprocessor and FPGA are
reduced.

References

1. *IP security protocol (ipsec)*, www.ietf.org/html.charters/ipsec-charter.html (2004).
2. *Transims: The transportation analysis simulation system*, http://www.transims.tsasa.lanl.gov/ (2004).
3. *International roadmap for semiconductors*, http://public.itrs.net/ (2005).
4. D. Abramson, A.d. Silva, M. Randall, and A. Posutla, *Special purpose computer architectures for high speed optimisation*, Second Australasian Conference on Parallel and Real Time Systems, 1995.
5. Actel Corporation, Mountain View, CA, *ProASIC Plus flash family FPGAs*, November 2004, Datasheet.
6. Jung Ho Ahn, William J. Dally, Brucek Khailany, and Ujval Kapasiand Abhishek Das, *Evaluating the imagine stream architecture*, Proceedings of the 31st International Symposium on Computer Architecture (2004).
7. Ali M. Al-Haj, *Fast discrete wavelet transformation using FPGAs and distributed arithmetic*, International Journal of Applied Science and Engineering **1** (2003), no. 2, 160–171.
8. O. T. Albaharna, P. Y. K. Cheung, and T. J. Clarke, *Area & time limitiations of FPGA-based virtual hardware*, Proceedings of the IEEE International Conference on Computer Design, October 1994, pp. 184–189.
9. Felix Albu, Jiri Kadlec, Chris Softley, Rudolf Matousek, Antonin Hermanek, Nick Coleman, and Anthony Fagan, *Implementation of (normalised) RLS lattice on Virtex*, Proceedings of 11th International Conference on Field-Programmable Logic and Applications (Belfast, Northern Ireland) (G. Brebner and R. Woods, eds.), Lecture Notes on Computer Science, vol. 2147, Springer-Verlag, August 2001, pp. 91–100.
10. Charles J. Alpert and Andrew B. Kahng, *Recent directions in netlist partitioning: a survey*, Integration, the VLSI Journal **19** (1995), no. 1-2, 1–81.
11. A. Alsolaim, J. Becker, M. Glesner, and J. Starzyk, *Architecture and application of a dynamically reconfigurable hardware array for future mobile communication systems*, Proceedings of the 2000 IEEE Symposium on Field-Programmable Custom Computing Machines (Napa, CA) (J. Arnold and K. Pocek, eds.), IEEE Computer Society, April 2000, pp. 205–214.
12. Altera Corporation, San Jose, CA, *SignalTap embedded logic analyzer Megafunction data sheet*, ver. 1.01 ed., January 2000.

13. Altera Corporation, San Jose, CA, *FLEX 10k embedded programmable logic device family*, version 4.2 ed., January 2003, Datasheet.

14. Altera Corporation, San Jose, CA, *APEX 20k programmable logic device family*, version 5.1 ed., March 2004, Datasheet.

15. Altera Corporation, San Jose, CA, *Stratix II device handbook, volume 1*, 1.2 ed., October 2004.

16. Altera Corporation, San Jose, CA, *Stratix II device handbook, volume 2*, 1.1 ed., July 2004.

17. S.F. Altschul, W. Gish, W. Miller, W.E. Myers, and D.J. Lipman, *Basic local alignment search tool*, Journal of Molecular Biology **215** (1990), no. 3, 403–410.

18. R. Amerson, R. Carter, B. Culbertson, P. Kuekes, and G. Snider, *Teramac-configurable custom computing*, Proceedings of IEEE Workshop on FPGAs for Custom Computing Machines (Napa, CA) (D. A. Buell and K. L. Pocek, eds.), April 1995, pp. 32–38.

19. R. Amerson, R. Carter, W. Culbertson, P. Kuekes, G. Snider, and Albertson L, *Plasma: An FPGA for million gate systems*, ACM/SIGDA International Symposium on Field Programmable Gate Arrays (Monterey, CA), February 1996.

20. Ray Andraka, *A survey of CORDIC algorithms for FPGA based computers*, FPGA '98: Proceedings of the 1998 ACM/SIGDA sixth international symposium on Field programmable gate arrays, ACM Press, 1998, pp. 191–200.

21. Annapolis Micro Systems, Annapolis, MD, *WILDFORCE reference manual*, rev 3.1 ed., 1998.

22. Annapolis Micro Systems, Inc., *http://www.annapmicro.com/*, (2004).

23. Andrew W. Appel and Maia Ginsburg, *Modern compiler implementation in C*, Cambridge University Press, 1998.

24. N. Aranki, W. Q. Jiang, and A. Ortega, *FPGA-based parallel implementation for the lifting discrete wavelet transform*, Proceedings of the SPIE - The International Society for Optical Engineering **4118** (2000), 96 – 107.

25. N. Aranki, A. Moopenn, and R. Tawel, *Parallel FPGA implementation of the split and merge discrete wavelet transform*, Lecture Notes In Computer Science **2438** (2002), 740 – 749.

26. J. M. Arnold, *The Splash 2 software environment*, Proceedings of IEEE Workshop on FPGAs for Custom Computing Machines (Napa, CA) (D. A. Buell and K. L. Pocek, eds.), April 1993, pp. 88–93.

27. M. Arnold, D.A. Buell, and E.G. Davis, *Splash 2*, 4th Annual Symposium on Parallel Algorithms and Architectures (San Diego, California), 1992, pp. 316–322.

28. Peter Ashenden, *The designer's guide to VHDL*, Academic Press, A Harcourt Science and Technology Company, 2002.

29. P. M. Athanas and H. F. Silverman, *Processor reconfiguration through instruction-set metamorphosis*, IEEE Computer **26** (1993), no. 3, 11–18.

30. K. A. Atkins, C. L. Barret, R. J. Beckman, S. G. Eubank, N. W. Hengarter, G. Istrate, A. V. S. Kumar, M. V. Marathe, H .S. Mortviet, C. M. Reidys, P. R. Romero, R. A. Pistone, J. P. Smith, P. E. Stretz, C. D. Engelhart, M. Droza, M. M. Morin, S. S. Pathak, S. Zust, and S. S. Ravi, *ADHOPNET: Integrated tools for end-to-end analysis of extremely large next generation telecommunication networks*, Tech. report, Los Alamos National Laboratory, Los Alamos, NM, 2003.

31. Navid Azizi, Ian Kuon, Aaron Egier, Ahmad Darabiha, and Paul Chow, *Reconfigurable molecular dynamics simulator*, IEEE International Symposium on FPGAs for Custom Computing Machines (2004), 197–206.

32. J. Babb, R. Tessier, and A. Agarwal, *Virtual wires: Overcoming pin limitations in FPGA-based logic emulators*, Proceedings of IEEE Workshop on FPGAs for Custom Computing Machines (Napa, CA) (D. A. Buell and K. L. Pocek, eds.), April 1993, pp. 142–151.

33. Sashisu Bajracharya, Chang Shu, Kris Gaj, and Tarek A. El-Ghazawi, *Implementation of elliptic curve cryptosystems over $GF(2^n)$ in optimal normal basis on a reconfigurable computer.*, FPL, 2004, pp. 1001–1005.

34. Z. K. Baker and V. K. Prasanna, *A methodology for synthesis of efficient intrusion detection systems on FPGAs*, Proceedings of IEEE Symposium on FPGAs for Custom Computing Machines (Napa, CA) (J. M. Arnold and K. L. Pocek, eds.), April 2004.

35. A. Banerjee, A. S. Dhar, and S. Banerjee, *FPGA realization of a CORDIC based FFT processor for biomedical signal processing*, Microprocessors and Microsystems **25** (2001), no. 3, 131 – 42.

36. V. Baumgarte, F. May, A. Nuckel, M. Vorbach, and M. Weinhardt, *Pact XPP— a self-reconfigurable data processing architecture*, Proceedings of the International Conference on Engineering of Reconfigurable Systems and Algorithms (Las Vegas, NV), CSREA Press, June 2001, pp. 64–70.

37. BEE 2, *http://bwrc.eecs.berkeley.edu/research/bee/bee2/*, (2004).

38. Pavle Belanović and Miriam Leeser, *A library of parameterized floating-point modules and their use*, FPL 2002: The 12th International Conference on Field-Programmable Logic and Applications, Springer-Verlag, 2002, pp. 657–666.

39. Peter Bellows, *SLAAC1-V SDK user's manual*, USC Information Sciences Institute—East, Arlingtonm, VA, 0.3.1 ed., October 2000, This document is included with the SLAAC1-V board documentation.

40. Peter Bellows, Jaroslav Flidr, Ladan Gharaj, et al., *IPSec-Protected Transport of HDTV over IP*, Military Applications of Programmable Logic Devices (MAPLD) (Wasington, DC) (Rich Katz, ed.), September 2003.

41. Peter Bellows, Jaroslav Flidr, Tom Lehman, Brian Schott, and Keith Underwood, *GRIP: A reconfigurable architecture for host-based gigabit-rate packet processing*, Proceedings of IEEE Symposium on FPGAs for Custom Computing Machines (Napa, CA) (J. M. Arnold and K. L. Pocek, eds.), April 2002.

42. Peter Bellows and Brad Hutchings, *JHDL-an HDL for reconfigurable systems*, Proceedings of the IEEE Symposium on Field-Programmable Custom Computing Machines (Napa, CA) (K. Pocek and J. Arnold, eds.), IEEE Computer Society, April 1998, pp. 175–184.

43. ———, *Designing run-time reconfigurable systems with JHDL*, Journal of VLSI Signal Processing **28** (2001).

44. A. Benkrid, K. Benkrid, and D. Crookes, *Design and implementation of a generic 2d orthogonal discrete wavelet transform on FPGA*, 11th Annual IEEE Symposium on Field-Programmable Custom Computing Machines. FCCM 2003, 9-11 April 2003, Napa, CA, USA, Los Alamitos, CA, USA : IEEE Comput. Soc, 2003, 2003, pp. 162 – 72.

45. K. Benkrid, D. Crookes, and A. Benkrid, *Towards a general framework for FPGA based image processing using hardware skeletons*, Parallel Computing **28** (2002).

46. D.A. Benson, I. Karsch-Mizrachi, D.J. Lipman, J. Ostell, and D.L. Wheeler, *Genbank: update*, Nucleic Acids Research, Database issue **32** (2004), D23–D26.

47. P. Bertin, D. Roncin, and J. Vuillemin, *Introduction to programmable active memories*, Systolic Array Processors (J. McCanny, J. McWhirther, and E. Swartslander Jr., eds.), Prentice Hall, 1989, pp. 300–309.

48. _____, *Programmable active memories: a performance assessment*, Research on Integrated Systems: Proceedings of the 1993 Symposium (G. Borriello and C. Ebeling, eds.), 1993, pp. 88–102.

49. Vaughn Betz and Jonathan Rose, *Automatic generation of FPGA routing architectures from high-level descriptions*, Proceedings of the 2000 ACM/SIGDA eighth international symposium on Field programmable gate arrays, ACM Press, 2000, pp. 175–184.

50. Vaughn Betz, Jonathan Rose, and Alexander Marquardt, *Architecture and CAD for deep-submicron FPGAs*, Kluwer Academic Publishers, Boston, 1999.

51. R. Bittner and P. Athanas, *Wormhole run-time reconfiguration*, ACM/SIGDA International Symposium on Field Programmable Gate Arrays (Monterey, CA), February 1997, pp. 79–85.

52. Per Bjureus, Mikael Millberg, and Axel Jantsch, *FPGA resource timing estimation from matlab execution traces*, International Conference on Signal Processing Applications and Technology (2002).

53. B.H. Bloom, *Space-time trade-offs in hash coding with allowable errors*, Commun. ACM **13** (1970), no. 7, 422–426.

54. W. Bohm, J. Hammes, B. Draper, M. Chawathe, C. Ross, and W. Najjar, *Mapping a single assignment programming language to reconfigurable systems*, Journal of Supercomputing **21** (2002), 117–130.

55. K. Bondalapati, P. Diniz, P. Duncan, J. Granacki, M. Hall, R. Jain, and H. Ziegler, *Defacto: Design environment for adaptive computing technology*, Proceedings of the Reconfigurable Architecture Workshop, held in conjunction with the International Parallel Processing Symposium, San Juan, Puerto Rico (1999).

56. Bernard Bosi, Guy Bois, and Yvon Savaria, *Reconfigurable pipelined 2-d convolvers for fast digital signal processing*, IEEE Transactions on Very Large Scale Integration (VLSI) Systems **7** (1999), no. 3.

57. Florian Braun, John Lockwood, and Marcel Waldvogel, *Protocol wrappers for layered network packet processing in reconfigurable hardware*, IEEE Micro **22** (2002), no. 3, 66–74.

58. S.E. Brenner, C. Chothia, and T.J. Hubbard, *Population statistics of protein structures: lessons from structural classification*, Curr. Opin. Struct. Biol **7** (1997), 369–376.

59. Stephen D. Brown, Robert J. Francis, Jonathan Rose, and Zvonko G. Vranesic, *Field-programmable gate arrays*, Kluwer Academic Publishers, 1992.

60. Duncan Buell, Jeffrey Arnold, and Walter Kleinfelder, *Splash 2: FPGAs in a custom computing machine*, Wiley- IEEE Computer Society Press, 1996.

61. Michael Caffrey, *A space-based reconfigurable radio*, Proceedings of the International Conference on Engineering of Reconfigurable Systems and Algorithms (ERSA) (Toomas P. Plaks and Peter M. Athanas, eds.), CSREA Press, June 2002, pp. 49–53.

62. Timothy J. Callahan, John R. Hauser, and John Wawrzynek, *The Garp architecture and C compiler*, IEEE Computer (2000).

63. Nick Camilleri, *Status and control semaphore registers using partial reconfigu-ration*, Application Note XAPP 153, Xilinx, Inc., San Jose, CA, June 1999.

64. A. Carreira, T. W. Fox, and L. E. Turner, *A method for implementing bit-serial finite impulse response digital filters in FPGAs using jbitssup TM*, Field-Programmable Logic and Applications. Reconfigurable Computing Is Going Mainstream. 12th International Conference, FPL 2002. Proceedings, 2-4 Sept. 2002, Montpellier, France (M. Glesner, P. Zipf, and M. Renovell, eds.), Berlin, Germany : Springer-Verlag, 2002, 2002, pp. 222 – 231.

65. L. Carter, R. Floyd, J. Gill, G. Markowsky, and M. Wegman, *Exact and ap-proximate membership testers*, 10th ACM Symposium on Theory of Computing (STOC 78), 1978, pp. 59–65.

66. Richard Carter, *Personal communication*, July 2001.

67. C. Chang, *BLAST implementation on BEE2*, University of California at Berke-ley, 2004.

68. Franois Charot, Eslam Yahya, and Charles Wagner, *Efficient modular-pipelined AES implementation in counter mode on ALTERA FPGA*, Field-Programmable Logic and Applications, vol. Lecture Notes in Computer Science vol. 2778, Springer, 2003, pp. 282 – 291.

69. D. C. Chen and J. M. Rabaey, *A reconfigurable multiprocessor IC for rapid prototyping of algorithmic-specific high-speed DSP data paths*, IEEE Journal of Solid-State Circuits **27** (1992), no. 12, 1895–1904.

70. Keping Chen, *Bit-serial realizations of a class of nonlinear filters based on pos-itive boolean functions*, IEEE Transactions on Circuits and Systems **36** (1989), no. 6.

71. Wen-Hsiung Chen, C. HarrisonSmith, and S. C. Fralick, *Fast computational algorithm for the discrete cosine transform*, IEEE Transactions on Communi-cations **COM-25** (1977), no. 9, 1004 – 1009.

72. Don Cherepacha and David Lewis, *A datapath oriented architecture for FPGAs*, Proceedings of the ACM/SIGDA International Symposium on Field Program-mable Gate Arrays (Monterey, CA), ACM/SIGDA, February 1994, pp. 1–11.

73. O. Y. H. Cheung, P. H. W. Leong, et al., *Implementation of an FPGA-based ac-celerator for virtual private networks*, IEEE International Conference on Field Programmable Technology, Springer, 2002.

74. O. Y. H. Cheung, K. H. Tsoi, K. H. Leung, et al., *Tradeoffs in parallel and serial implementations of the International Data Encryption Algorithm IDEA*, Proceedings of the Cryptographic Hardware and Embedded Systems Workshop (CHES), Springer, 2001, pp. 333–347.

75. Kang-Ngee Chia, Hea Joung Kim, Shane Lansing, William H. Mangione-Smith, and John Villasenor, *High-performance automatic target recognition through data-specific VLSI*, IEEE Transactions on Very Large Scale Integration (VLSI) Systems **6** (1998), no. 3.

76. Y. H. Cho and W. H. Mangione-Smith, *Deep packet filter with dedicated logic and read only memories*, Proceedings of IEEE Symposium on FPGAs for Cus-tom Computing Machines (Napa, CA) (J. M. Arnold and K. L. Pocek, eds.), April 2004.

77. Seonil Choi and Viktor Prasanna, *Time and energy efficient matrix factoriza-tion using FPGAs*, FPL 03: 13th International Conference on Field Program-mable Logic and Applications, September 2003.

78. N. Chomsky, *Syntactic structures*, Mouton, La Haye, 1957.

79. C. Chou, S. Mohanakrishnan, and J. B. Evans, *FPGA implementation of digital filters*, Proceedings of the Fourth International Conference on Signal Processing Applications and Technology (Santa Clara, CA), 1993, pp. 80–88.

80. E. Chow, T. Hunkapiller, and J. Peterson, *Biological information signal processor*, ASAP'91, International Conference on Application Specific Array Processors (Barcelona, Spain), 1991.

81. S. Churcher, T. Kean, and B. Wilkie, *The XC6200 fastmap^{TM} processor interface*, Field-Programmable Logic and Applications : 5th international workshop (Oxford, United Kingdom) (W. Moore and W. Luk, eds.), Lecture Notes in Computer Science, vol. 975, Springer-Verlag, Berlin, August/September 1995, pp. 36–43.

82. C. Ciressan, *An FPGA-based syntactic parser for real-life context free grammars*, Ph.D. thesis, EPFL, january 2002.

83. C. R. Clark and D. E. Schimmel, *Scalable pattern matching for high speed networks*, Proceedings of IEEE Symposium on FPGAs for Custom Computing Machines (Napa, CA) (J. M. Arnold and K. L. Pocek, eds.), April 2004.

84. Netscape Communications, *SSL 3.0 Specification*, wp.netscape.com/eng/ssl3 (1996).

85. Katherine Compton and Scott Hauck, *Reconfigurable computing: a survey of systems and software*, ACM Computing Surveys (CSUR) (2002), no. 2, 171–210.

86. _____, *Flexibility measurement of domain-specific reconfigurable hardware*, FPGA '04: Proceeding of the 2004 ACM/SIGDA 12th international symposium on Field programmable gate arrays, ACM Press, 2004, pp. 155–161.

87. J. Cong and Y. Ding, *FlowMap: An optimal technology mapping algorithm for delay optimization in lookup-table based FPGA designs*, IEEE Transactions on Computer-Aided Design of Integrated Circuits and Systems **13** (1994), no. 1, 1–11.

88. _____, *On area/depth trade-off in LUT-based FPGA technology mapping*, IEEE Transactions on VLSI Systems **2** (1994), no. 2, 137–148.

89. Jason Cong, Honching Peter Li, Sung Kyu Lim, Toshiyuki Shibuya, and Dongmin Xu, *Large scale circuit partitioning with loose/stable net removal and signal flow based clustering*, ICCAD '97: Proceedings of the 1997 IEEE/ACM international conference on Computer-aided design, IEEE Computer Society, 1997, pp. 441–446.

90. Jason Cong and Sung Kyu Lim, *Edge separability based circuit clustering with application to circuit partitioning*, ASP-DAC '00: Proceedings of the 2000 conference on Asia South Pacific design automation, ACM Press, 2000, pp. 429–434.

91. Jason Cong and M'Lissa Smith, *A parallel bottom-up clustering algorithm with applications to circuit partitioning in VLSI design*, DAC '93: Proceedings of the 30th international conference on Design automation, ACM Press, 1993, pp. 755–760.

92. G. A. Constantinides, *Perturbation analysis for word-length optimization*, 11th Annual IEEE Symposium on Field-Programmable Custom Computing Machines. FCCM 2003, 9-11 April 2003, Napa, CA, USA, Los Alamitos, CA, USA : IEEE Comput. Soc, 2003, 2003, pp. 81 – 90.

93. A.A. Conti, T. Van Court, and M.C. Herbordt, *Processing repetitive sequence structures with mismatches at streaming rate.*, FPL 2004, 2004, pp. 1080–1083.

94. J. W. Cooley and J. W. Tukey, *An algorithm for the machine calculation of complex fourier series*, Math. Computation **19** (1965), 297–301.

95. F. Corpet, F. Serav, J. Gouzy, and D. Kahn, *ProDom and ProDom-CG: tools for protein domain analysis and whole genome comparisons*, Nucleic Acid Research **28** (2000), 267–269.

96. T. Van Court and M. C. Herbordt, *Families of FPGA-based algorithms for approximate string matching*, ASAP'04, 2004.

97. Tom Van Court, Yongfeng Gu, and Martin C. Herbordt, *Fpga acceleration of rigid molecule interactions*, IEEE International Symposium on FPGAs for Custom Computing Machines (2002), 197–206.

98. M. Crochemore, C. Iliopoulos, Y. Pinzon, and J. Reid, *A fast and practical bit-vector algorithm for the longest common subsequence problem*, Information Processing Letters **80** (2001), no. 6, 279–285.

99. C. Ebeling D. C. Cronquist and P. Franklin, *RaPiD – reconfigurable pipelined datapath*, Field-Programmable Logic: Smart Applications, New Paradigms, and Compilers. 6th International Workshop on Field-Programmable Logic and Applications (Darmstadt, Germany) (R. W. Hartenstein and M. Glesner, eds.), Springer-Verlag, September 1996, pp. 126–135.

100. D. C. Cronquist, C. Fisher, M. Figueroa, P. Franklin, and C. Ebeling, *Architecture design of reconfigurable pipelined datapaths*, Proceedings of the 20th Anniversary Conference on Advanced Research in VLSI (Atlanta, GA), March 1999, pp. 23–40.

101. Darren C. Cronquist, Paul Franklin, Stefan G. Berg, and Carl Ebeling, *Specifying and compiling applications for RaPiD*, Proceedings. IEEE Symposium on FPGAs for Custom Computing Machines (Napa, CA) (K. Pocek and J. Arnold, eds.), IEEE Computer Society, April 1998, pp. 116–125.

102. Joan Daemen and Vincent Rijmen, *The design of Rijndael: AES — the Advanced Encryption Standard*, Springer-Verlag, 2002.

103. Andread Dandalin, Viktor Prasanna, and Jose Rolim, *An adaptive cryptographic engine for IPSec architectures*, Proceedings of IEEE Symposium on FPGAs for Custom Computing Machines (Napa, CA) (Brad Hutchings, ed.), April 2000.

104. A. Darling, L. Carey, and W. Feng, *The design, implementation, and evaluation of mpiBLAST*, CWCE 2003, 2003.

105. D. Dasgupta and Z. Michalewicz, *Evolutionary algorithms in engineering applications*, Springer-Verlag, Berlin, 1997.

106. Janaka Deepakumara, Howard M. Heys, and R. Venkatesan, *FPGA implementation of md5 hash algorithm*, IEEE Canadian Conference on Electrical and Computer Engineering, May 2001.

107. A. DeHon, *DPGA-coupled microprocessors: Commodity ICs for the early 21st century*, Proceedings of IEEE Workshop on FPGAs for Custom Computing Machines (Napa, CA) (D. A. Buell and K. L. Pocek, eds.), April 1994, pp. 31–39.

108. A. DeHon, J. Adams, M. DeLorimier, N. Kapre, Y. Matsuda, and H. Naeimi, *Design patterns for reconfigurable computing*, IEEE International Symposium on FPGAs for Custom Computing Machines (2004), 13–22.

109. André DeHon, *DPGA utilization and application*, ACM/SIGDA International Symposium on Field Programmable Gate Arrays (Monterey, CA), February 1996, pp. 115–121.

110. J. P. Delahaye, G. Gogniat, C. Roland, and P. Bomel, *Software radio and dynamic reconfiguration on a DSP/FPGA platform*, Frequenz **58** (2004), no. 5-6, 152 – 159.

111. P. B. Denyer and David Renshaw, *VLSI signal processing; a bit-serial approach*, Addison-Wesley Longman Publishing Co., Inc., 1985.

112. T. Desai and K. J. Hintz, *A parallel implementation of LMS adaptive filter in hardware for landmine detection*, Proceedings of SPIE - The International Society for Optical Engineering **5415** (2004), no. PART 2, 973 – 983.

113. Jérémie Detrey and Florent de Dinechin, *FPLibrary, a VHDL library of parametrisable floating-point and LNS operators for FPGA*, http://perso.ens-lyon.fr/jeremie.detrey/FPLibrary/, 2004.

114. S. Dharmapurikar, M. Attig, and J. Lockwood, *Design and implementation of a string matching system for network intrusion detection using FPGA-based Bloom filters*, IEEE Symposium on Field-Programmable Custom Computing Machines (FCCM'04), 2004.

115. S. Dharmapurikar, P. Krishnamurthy, T. Sproull, and J. Lockwood, *Deep packet inspection using parallel bloom filters*, 2003.

116. C. Dick, *Implementing area optimized narrow-band FIR filters using Xilinx FPGAs*, Proceedings of the SPIE - The International Society for Optical Engineering **3526** (1998), 227 – 38.

117. _____, *Minimum multiplicative complexity implementation of the 2-D DCT using Xilinx FPGAs*, Proceedings of the SPIE - The International Society for Optical Engineering **3526** (1998), 190 – 201.

118. C. Dick and F. Harris, *On the use of FPGAs for OFDM signal processing*, International Conference on Engineering of Reconfigurable Systems and Algorithms (ERSA 04) (Las Vegas, NV) (TP Plaks, ed.), CSREA PRESS, June 2004, pp. 259 – 263.

119. C. Dick, F. Harris, and M. Rice, *Synchronization in software radios - carrier and timing recovery using FPGAs*, Proceedings of the IEEE Symposium on Field-Programmable Custom Computing Machines (Napa, CA) (B. L. Hutchings, ed.), IEEE COMPUTER SOC, April 2000, pp. 195 – 204.

120. C. H. Dick, *FPGAs: re-inventing the signal processor*, International Conference on Engineering of Reconfigurable Systems and Algorithms - ERSA'03, 23-26 June 2003, Las Vegas, NV, USA (TP Plaks, ed.), Athens, GA, USA : CSREA Press, 2003, 2003, pp. 252 – 8.

121. Chris Dick, *Computing the discrete fourier transform on FPGA based systolic arrays*, ACM/SIGDA International Symposium on Field Programmable Gate Arrays (Monterey, CA), February 1996, pp. 129–135.

122. Chris Dick, Fred Harris, Fred, and Michael Rice, *FPGA implementation of carrier synchronization for QAM receivers*, Journal of VLSI Signal Processing **36** (2004), no. 1, 57–71.

123. E. W. Dijkstra, *A note on two problems in connexion with graphs*, Numer. Math. **1** (1959), 269–271. MR MR0107609 (21 #6334)

124. Tom Dillon, *Two Virtex-II FPGAs deliver fastest, cheapest, best high-performance image processing system*, Xcell Journal (2001), no. Fall/Winter, 70–73, Xilinx, Inc. publication.

125. _____, *Personal communications*, May 2004.

126. T. T. Do, H. Kropp, C. Reuter, and P. Pirsch, *A flexible implementation of high-performance FIR filters on Xilinx FPGAs*, Field-Programmable Logic and

Applications. From FPGAs to Computing Paradigm. 8th International Workshop, FPL '98. Proceedigns, 31 Aug.-3 Sept. 1998, Tallinn, Estonia (A. Hartenstein, RW; Keevallik, ed.), Berlin, Germany : Springer-Verlag, 1998, 1998, pp. 441 – 5.

127. Bruce A. Draper, Ross Beveridge, A.P. Willem Böhm, Charles Ross, and Monica Chawathe, *Accelerated image processing on FPGAs*, IEEE Transactions on Image Processing **12** (2003), no. 12.

128. DRS Technologies, Inc., *http://www.drs.com/*, (2004).

129. J. Dumoulin, J.A. Foster, J.F. Frenzel, and S. McGrew, *Special purpose image convolution with evolvable hardware*, Workshop on Real-World Applications of Evolutionary Computing, Lecture Notes in Computer Science, vol. 1803, 2000.

130. P. Dunn and P. Corke, *Real-time stereopsis using FPGAs*, 7th International Workshop on Field-Programmable Logic and Applications: FPL '97, September 1997.

131. Lisa Durbeck and Nick Macias, *The cell matrix: An architecture for nanocomputing*, Nanotechnology (2001), 217–230.

132. S. Dydel and P. Bala, *Large scale protein sequence alignment using FPGA reprogrammable logic devices*, FPL'04, LNCS 3203 (Springer, ed.), 2004, pp. 23–32.

133. C. Ebeling, L. McMurchie, S. A. Hauck, and S. Burns, *Placement and routing tools for the Triptych FPGA*, IEEE Transactions on VLSI Systems **3** (1995), no. 4, 473–482.

134. Carl Ebeling, Darren C. Cronquist, Paul Franklin, Jason Secosky, and Stefan G. Berg, *Mapping applications to the RaPiD configurable architecture*, Proceedings of the 5th Annual IEEE Symposium on FPGAs for Custom Computing Machines (Napa, CA) (K. Pocek and J. Arnold, eds.), IEEE Computer Society, April 1997, pp. 106–115.

135. S.R. Eddy, *Profil hidden Markov model*, Bioinformatics **14** (2002), 755–763.

136. A. J. Elbirt and C. Paar, *An FPGA implementation and performance evaluation of the Serpent block cipher*, FPGA '00: Proceedings of the 2000 ACM/SIGDA eighth international symposium on Field Programmable Gate Arrays, ACM Press, 2000, pp. 33–40.

137. Rolf Enzler, Tobias Jeger, Didier Cottet, and Gerhard Troster, *High-level area and performance estimation of hardware building blocks on FPGAs*, International Conference on Field-Programmable Logic and Applications, (2000), 525–534.

138. Eonic, *Eonic PowerFFT on-line datasheet*, http://www.eonic.com/cgi-bin/update_frames.cgi?lang=uk&go=processors/powerfft.html, April 2004.

139. G. Estrin, *Organization of computer systems – the fixed plus variable structure computer*, Proceedings of the Western Joint Computer Conference, 1960, pp. 33–40.

140. G. Estrin, B. Bussell, R. Turn, and J. Bibb, *Parallel processing in a restructurable computer system*, IEEE Transactions on Electronic Computers (1963), 747–755.

141. ———, *Parallel processing in a restructurable computer system*, IEEE Transactions on Electronic Computers (1963), 747–755.

142. Gerald Estrin, *Reconfigurable computer origins: The UCLA fixed-plus-variable (F+V) structure computer*, IEEE Annals of the History of Computing **24** (2002), no. 4, 3–9.

143. A. Bateman et al., *The PFAM protein families database*, Nucleic Acids Research, Database Issue **32** (2004), D138–D141.

144. B. Boeckmann et al., *The swiss-prot protein knowledgebase and its supplement TrEMBL in 2003*, Nucleic Acids Research **31** (2003), no. 1, 365–370.

145. F. Sanger et al, *Nucleotide sequence of bacteriophage lambda DNA*, J. Molecular Biology **162** (1982), no. 4, 729–773.

146. G. Stoesser et al., *The EMBL nucleotide sequence database*, Nucleic Acids Research **30** (2002), 21–26.

147. R.K. Singh et al., *Research on integrated system*, ch. A Scalable Systolic Multiprocessor System dor Analysis of Biological Sequences, G. Borrielo and C. Ebeling, 1993.

148. S.F. Altschul et al., *Gapped BLAST and PSI-BLAST: a new generation of protein database search programs*, Nucleic Acids Research **25** (1997), no. 17, 3389–3402.

149. S. Eubank, H. Guclu, V. S. A. Kumar, M. V.Madhav, A. Srinivasan, Z. Toroczkai, and N. Wang, *Modelling disease outbreaks in realistic urban social networks*, Nature **429** (2004), no. 6988, 180–184.

150. J. B. Evans, *Efficient FIR filter architectures suitable for FPGA implementation*, IEEE Transactions on Circuits and Systems II: Analog and Digital Signal Processing **41** (1994), no. 7, 490 – 3.

151. Joe Fabula, Austin Lesea, and Saar Drimer, *The NSEU sensitivity of static latch based FPGAs and flash storage devices*, Proceedings of the 7th Military and Aerospace Applications of Programmable Logic Devices (Washington, D.C.) (Richard Katz, ed.), NASA Office of Logic Design, AIAA, September 2004, On-line proceedings available through http://www.klabs.org., pp. L139.1–26.

152. Olivier Faugeras, Thierry Vieville, E. Theron, J. Vuillemin, B. Hotz, Z. Zhang, L. Moll, P. Bertin, H. Mathieu, P. Fua amd G. Berry, and C. Proy, *Real-time correlation-based stereo: Algorithm, implementations and applications*, Technical Report 2013, INRIA, 1993.

153. B. Feher, *New inner product algorithm of the two-dimensional DCT*, Proceedings of the SPIE - The International Society for Optical Engineering **2419** (1995), 436 – 44.

154. J.P. Fitch, E.J. Coyle, and N.C. Gallagher, *Median filtering by threshold decomposition*, IEEE Trans. Acoust., Speech, Signal Processing **ASSP-32** (1984), no. 6.

155. T. W. Fox and L. E. Turner, *Implementing the discrete cosine transform using the Xilinx Virtex FPGA*, Lecture Notes In Computer Science **2438** (2002), 492 – 502.

156. Robert Francis, *A tutorial on logic synthesis for lookup-table based FPGAs*, IEEE International Conference on Computer Aided Design (1992), 40–47.

157. Robert Francis, Jonathan Rose, and Zvonko Vranesic, *Chortle-crf: Fast technology mapping for lookup table-based FPGAs*, DAC '91: Proceedings of the 28th conference on ACM/IEEE design automation, ACM Press, 1991, pp. 227–233.

158. Robert J. Francis, Jonathan Rose, and Kevin Chung, *Chortle: a technology mapping program for lookup table-based field programmable gate arrays*, DAC '90: Proceedings of the 27th ACM/IEEE conference on Design automation, ACM Press, 1990, pp. 613–619.

159. Jan Frigo, Tom Braun, Joe Arrowood, and Maya Gokhale, *Comparison of high-level FPGA design tools for a bpsk signal detection application*, Software Defined Radio Forum (2003).

160. B. K. Fross, R. L. Donaldson, and D. J. Palmer, *PCI-based WILDFIRE reconfigurable computing engines*, Proceedings of SPIE—The International Society for Optical Engineering (Bellingham, WA), vol. 2914, SPIE, SPIE, November 1996, pp. 170–179.

161. Earl Fuller, Michael Caffrey, Phil Blain, Carl Carmichael, Noor Khalsa, and Anthony Salazar, *Radiation test results of the Virtex FPGA and ZBT SRAM for space based reconfigurable computing*, Proceedings of the Military and Aerospace Programmable Logic Devices International Conference(MAPLD) (Laurel, MD), September 1999.

162. F. Furtek, *A field-programmable gate array for systolic computing*, Research on Integrated Systems: Proceedings of the 1993 Symposium (G. Borriello and C. Ebeling, eds.), 1993, pp. 183–199.

163. Kris Gaj and Pawel Chodowiec, *Fast implementation and fair comparison of the final candidates for advanced encryption standard using field programmable gate arrays*, Progress in Cryptology - CT-RSA 2001: The Cryptographers' Track at RSA Conference, vol. Lecture Notes in Computer Science vol. 2020, Springer, 2001.

164. Lijun Gao, K. K. Parhi, and Jun Ma, *Relaxed annihilation-reordering look-ahead QRD-RLS adaptive filters*, Journal of VLSI Signal Processing Systems for Signal, Image, and Video Technology **35** (2003), no. 2, 119 – 35.

165. P. Gardner-Stephen and G. Knowles, *Dash: Localising dynmamic programming for order of magnitude faster, accurate sequence alignment*, IEEE Computational Systems Bioinformatics Conference (San Francisco, CA), 2004.

166. R.A. Georges and J. Heringa, *Protein domain identification and improved sequence similarity searching using PSI-BLAST*, Proteins: Structure, Function, and Genetics **48** (2002), 672–681.

167. M. Giraud and D. Lavenier, *Linear encoding scheme for weighted finite automata*, Ninth International Conference on Implementation and Application of Automata (CIAA 2004), LNCS 3317, july 2004.

168. M. Gokhale, W. Holmes, A. Kopser, S. Lucas, R. Minnich, D. Sweely, and D. Lopresti, *Building and using a highly parallel programmable logic array*, IEEE Computer **24** (1991), no. 1, 81–89.

169. M. Gokhale and B. Schott, *Data parallel C on a reconfigurable logic array*, Journal of Supercomputing **9** (1994), no. 3, 291–313.

170. M. B. Gokhale and J. M. Stone, *Co-synthesis to a hybrid RISC/FPGA architecture*, Journal of VLSI Signal Processing Systems **24** (2000).

171. M. B. Gokhale, J. M. Stone, J. Arnold, and M. Kalinowski, *Stream-oriented FPGA computing in the Streams-C high level language*, IEEE International Symposium on FPGAs for Custom Computing Machines, 2000.

172. Maya Gokhale, Dave Dubois, Andy Dubois, Mike Boorman, Steve Poole, and Vic Hogsett, *Granidt: Towards gigabit rate network intrusion detection technology.*, FPL, 2002, pp. 404–413.

173. Maya Gokhale and Janice Stone, *NAPA C: compiling for a hybrid RISC/FPGA archictecture*, Proceedings of IEEE Symposium on FPGAs for Custom Computing Machines (Napa, CA) (J. M. Arnold and K. L. Pocek, eds.), April 1997.

174. S. C. Goldstein, H. Schmit, M. Budiu, S. Cadambi, M. Moe, and R. R. Taylor, *PipeRench: a reconfigurable architecture and compiler*, IEEE Computer **33** (2000), no. 4, 70–77.

175. Ivan Gonzalez, Sergio López-Buedo, Francisco J. Gómez, and Javier Martínez, *Using partial reconfiguration in cryptographic applications: An implementation of the idea algorithm.*, FPL, 2003, pp. 194–203.

176. John B. Gosling, *Simulation in the design of digital electronic systems*, Cambridge University Press, 1993.

177. O. Gotoh, *An improved algorithm for matching biological sequences*, J. Mol. Biol. **162** (1982), no. 3, 705–708.

178. J. Gouzy, F. Corpet, and D. Kahn, *Whole genome protein domain analysis using a new method for domain clustering*, Computers and Chemistry (1999), no. 23, 333–340.

179. Paul Graham and Brent Nelson, *FPGA-based sonar processing*, ACM/SIGDA International Symposium on Field Programmable Gate Arrays - FPGA (1998), 201 – 208.

180. _____, *FPGAs and DSPs for sonar processing—inner loop computations*, Technical Report CCL-1998-GN-1, Brigham Young University, Provo, UT, 1998, Published on http://splish.ee.byu.edu.

181. Paul Graham, Brent Nelson, and Brad Hutchings, *Instrumenting bitstreams for debugging FPGA circuits*, Proceedings of the 9th Annual IEEE Symposium on Field-Programmable Custom Computing Machines (Rohnert Park, CA), IEEE Computer Society, April 2001.

182. Paul S. Graham, *Logical hardware debuggers for FPGA-based systems*, Ph.D. thesis, Brigham Young University, Provo, UT, December 2001.

183. L. Grate, M. Diekhans, D. Dahle, and R. Hughey, *Sequence analysis with the kestrel SIMD parallel processor*, Pacific Symposium on Biocomputing (Hawaii), 2001.

184. A. A. Gray, C. Lee, P. Arabshahi, and J. Srinivasan, *Object-oriented reconfigurable processing for wireless networks*, IEEE International Conference on Communications, 28 April-2 May 2002, New York, NY, USA, Piscataway, NJ, USA : IEEE, 2002, 2002, pp. 497 – 501 vol.1.

185. S. Guccione and E. Keller, *Gene matching using JBits*, Field Programmable Logic and Applications, Reconfigurable Comuting 12th Int. Conference, 2002.

186. Steven A. Guccione, Delon Levi, and Prasanna Sundararajan, *JBits: A Java-based interface for reconfigurable computing*, Second Annual Military and Aerospace Applications of Programmable Devices and Technologies Conference (MAPLD) (Laurel, MD), September 1999.

187. P. Guerdoux and D. Lavenier, *Samba: Hardware accelerator for biological sequence comparison*, CABIOS **13** (1997), no. 6, 609–615.

188. P. Guerdoux-Jamet and D. Lavenier, *Systolic filter for fast DNA similarity search*, ASAP'95, International Conference on Application Specific Array Processors (Strasbourg, France), 1995.

189. S. Gupta, *Hardware acceleration of hidden markov models for bioinformatics applications*, Master's thesis, Boise State University, 2004.

190. S. Guyetant, M. Giraud, S. Derrien, L. L'Hours, S. Rubini, F. Raimbault, and D. Lavenier, *Cluster of reconfigurable nodes for scanning large genomic banks*, Parallel Computing **31** (2005), no. 1.

191. Lars Hagen and Andrew B. Kahng, *A new approach to effective circuit clustering*, ICCAD '92: Proceedings of the 1992 IEEE/ACM international conference on Computer-aided design, IEEE Computer Society Press, 1992, pp. 422–427.

192. Malay Haldar, Anshuman Nayak, Alok Choudhary, and Prith Banerjee, *A system for synthesizing optimized FPGA hardware from MATLAB*, Proceedings of the 2001 IEEE/ACM International Conference on Computer-aided design (2001), 314–319.

193. Tsuyoshi Hamada, Toshiyuki Fukushige, Atsushi Kawai, and Joshiyuki Makino, *Progrape-1: A programmable special-purpose computer for many-body simulations*, IEEE International Symposium on FPGAs for Custom Computing Machines (1998), 256–257.

194. P. Hamalainen et al., *Hardware implementation of the improved WEP and RC4 encryption algorithms for wireless terminals*, European Signal Processing Conference (EUSIPCO '2000), September 2000.

195. T. Han and S. Parameswaran, *Swasad: An ASIC design for high speed DNA sequence matching*, 15th Int. Conf. on VLSI Design (IEEE CSP, ed.), 2002.

196. Silvina Zimi Hanono, *Innerview hardware debugger: A logic analysis tool for the virtual wires emulation system*, Master's thesis, Massachusetts Institute of Technology, February 1995.

197. R. Hartenstein, *A decade of reconfigurable computing: a visionary retrospective*, DATE '01: Proceedings of the conference on Design, automation and test in Europe, IEEE Press, 2001, pp. 642–649.

198. R. Hartenstein, R. Kress, and H. Reinig, *A new FPGA architecture for word-oriented datapaths*, Field-Programmable Logic: Architectures, Synthesis and Applications. 4th International Workshop on Field-Programmable Logic and Applications (Prague, Czech Republic) (R. Hartenstein and M. Z. Servit, eds.), Springer-Verlag, September 1994, pp. 144–155.

199. R. I. Hartley and K. K. Parhi, *Digit-serial computation*, Kluwer Academic Press, Boston, MA, 1995.

200. S. Haruyama, R. Morelos-Zaragoza, and Y. Sanada, *A software defined radio platform with direct conversion: SOPRANO*, Wireless Personal Communications **23** (2002), no. 1, 67 – 76.

201. John R. Hauser and John Wawrzynek, *GARP: A MIPS processor with a reconfigurable coprocessor*, Proceedings of IEEE Workshop on FPGAs for Custom Computing Machines (Napa, CA) (J. Arnold and K. L. Pocek, eds.), April 1997.

202. D. Haussler, A. Krogh, I.S. Mian, and K. Sjolander, *Protein modelling using hidden markov models: Analysis of globins*, Hawaii Int. Conf. System Sciences, January 1993.

203. Dirk Helgemo, *Digital signal processing at 1ghz in a field-programmable object array*, Proceedings of 6th Annual Military and Aerospace Programmable Logic Devices International Conference (Washington, D.C.), NASA Office of Logic Design, September 2003, http://www.klabs.org, pp. D1.1–5.

204. K. S. Hemmert and B. Hutchings, *Issues in debugging highly parallel FPGA-based applications derived from source code.*, Proceedings of the Asia and South Pacific Design Automation Conference (Kitakyushu, Japan), Piscataway, NJ, USA : IEEE, 2003, January 2003, pp. 483 – 488.

205. K. Scott Hemmert, Justin L. Tripp, Brad L. Hutchings, and Preston A. Jackson, *Source level debugger for the sea cucumber synthesizing compiler*, Proceed-

ings of the 11th Annual IEEE Symposium on Field-Programmable Custom Computing Machines, IEEE Computer Society, April 2003, pp. 228–237.

206. J.G. Henikoff and S. Henikoff, *Amino acid substitution matrices from protein blocks*, Proc. Natl. Acad. Sci. USA **89** (1992), 10915–10919.

207. J. L. Hennessy and D. A. Patterson, *Computer architecture: A quantitative approach, 3rd edition*, Morgan Kaufmann Publishing Co., 2003.

208. Hewlett Packard, Palo Alto, CA, *Tmac*, Aug 1995, This is documentation which came with Teramac held in the BYU Configurable Computing Lab's document library. Apparently, this manual was a chapter in some internal document created by HP Labs, but the name of the document is not known.

209. Todd Hiers, *Tms320c6414/5/6 power consumption summary*, Application Report SPRA811C, Texas Instruments, July 2003.

210. D.T. Hoang, *Searching genetic databases on splash 2*, FCCM'93, IEEE Workshop on FPGAs for Custom Computing Machines (Napa, California), 1993, pp. 185–191.

211. Alireza Hodjat and Ingrid Verbauwhede, *A 21.54 gbits/s fully pipelined AES processor on FPGA*, Proceedings of IEEE Symposium on FPGAs for Custom Computing Machines (Napa, CA) (J. M. Arnold and K. L. Pocek, eds.), April 2004.

212. E. B. Hogenauer, *An economical class of digital filters for decimation and interpolation*, IEEE Transactions on Acoustics, Speech and Signal Processing **ASSP/29** (1981), no. 2, 155 – 62.

213. M. Honda, H. Harada, and M. Fujise, *An efficient radio transmission scheme of configuration data for FPGA-based downloadable software radio communication systems*, Electronics and Communications in Japan, Part 1 (Communications) **86** (2003), no. 7, 31 – 41.

214. J. Hopcroft and J. Ullman, *Formal languages and their relation to automata*, Addison-Wesley, 1969.

215. N. Howard and R. W. Taylor, *Reconfigurable logic: technology and applications*, Computing & Control Engineering Journal **3** (1992), no. 5, 235–240.

216. Shen-Fu Hsiao, Chun-Yi Lau, and J. M. Delosme, *Redundant constant-factor implementation of multi-dimensional CORDIC and its application to complex SVD*, Journal of VLSI Signal Processing Systems for Signal, Image, and Video Technology **25** (2000), no. 2, 155 – 66.

217. Y. H. Hu, *CORDIC-based VLSI architectures for digital signal processing*, IEEE Signal Processing Magazine **9** (1992), no. 3, 16 – 35.

218. R. Hughey, *Massively parallel biosequence analysis*, Tech. Report UCSC-CRL-93-14, University of California, Santa Cruz, 1993.

219. _____, *Parallel hardware for sequence comparison alignement*, CABIOS **12** (1996), no. 6, 473–479.

220. N. Hulo, C.J. Sigrist, V. Le Saux, P.S. Langendijk-Genevaux, L. Bordoli, A. Gattiker, E. De Castro, P. Bucher, and A. Bairoch, *Recent improvements to the PROSITE database*, Nucl. Acids. Res. **32** (2004), D134–D137.

221. B. Hutchings, P. Bellows, J. Hawkins, S. Hemmert, B. Nelson, and M. Rytting, *A CAD suite for high-performance FPGA design*, Proceedings of the Seventh Annual IEEE Symposium on Field-Programmable Custom Computing Machines (Napa, CA) (K. Pocek and J. Arnold, eds.), IEEE Computer Society, April 1999, pp. 12–24.

222. B. L. Hutchings, R. Franklin, and D. Carver, *Assisting network intrusion detection with reconfigurable hardware*, Proceedings of IEEE Symposium on FPGAs for Custom Computing Machines (Napa, CA) (J. M. Arnold and K. L. Pocek, eds.), April 2002.

223. B. L. Hutchings and B. E. Nelson, *Unifying simulation and execution in a design environment for FPGA systems*, IEEE Transactions on Very Large Scale Integration (VLSI) Systems **9** (2001), no. 1, 201–205.

224. Brad L. Hutchings and Brent E. Nelson, *Gigaop DSP on FPGA*, Journal of VLSI Signal Processing Systems **36** (2004), no. 1, 41–55.

225. Jeng-Kuang Hwang and Cha-Hsing Chu, *FPGA implementation of an all-digital T/2-spaced QPSK receiver with Farrow interpolation timing synchronizer and recursive Costas loop*, 2004 IEEE Asia-Pacific Conference on Advanced System Integrated Circuits, 4-5 Aug. 2004, Fukuoka, Japan, Piscataway, NJ, USA : IEEE, 2004, 2004, pp. 248 – 51.

226. Heungjae Im, Seungheon Hyeon, Weon-Cheol Lee, Hwanseog Bahk, Cheolhoon Lee, Jonghun Kim, and Seungwon Choi, *Implementation of smart antenna base station for IS-2000 1X*, IEEE Vehicular Technology Conference **57** (2003), no. 1, 582 – 586.

227. Annapolis Micro Systems Inc., *Firebird reconfigurable computer*, http://www.annapmicro.com, 2004.

228. M. Iwata, I. Kajitani, H. Yamada, H. Iba, and T. Higuchi, *A pattern recognition system using evolvable hardware*, Proceedings of Parallel Problem Solving from Nature IV - PPSN IV, 1996.

229. Anil K. Jain, *Fundamentals of digital image processing*, Information and System Sciences Series, Prentice-Hall, Englewood Cliffs, NJ, 1989.

230. Hongtu Jiang and V. Owall, *Fpga implementation of real-time image convolutions with three level of memory hierarchy*, IEEE International Conference on Field-Programmable Technology (FPT), December 2003.

231. Don H. Johnson and Dan E. Dudgeon, *Array signal processing: Concepts and techniques*, Prentice Hall Signal Processing Series, Prentice-Hall, Englewood Cliffs, NJ, 1993.

232. K.-C.Chen, J. Cong, Y. Ding, A. B. Kahng, and P. Trajmar, *DAG-Map: Graph-based FPGA technology mapping for delay optimization*, IEEE Design and Test of Computers **9** (1992), no. 3, 7–20.

233. Chi-Chou Kao and Yen-Tai Lai, *An efficient algorithm for finding the minimal-area FPGA technology mapping*, ACM Trans. Des. Autom. Electron. Syst. **10** (2005), no. 1, 168–186.

234. Thomas A. Kean, *Configurable logic: A dynamically programmable cellular architecture and its VLSI implementation*, Ph.D. thesis, University of Edinburgh, 1988.

235. Eric Keller, *JRoute: A run-time routing API for FPGA hardware*, Parallel and Distributed Processing: 15 IPDPS 2000 Workshops (Cancun, Mexico) (José Romlin et al., eds.), Lecture Notes in Computer Science, vol. 1800, Springer Verlag, May 2000, Seventh Reconfigurable Architectures Workshop (RAW 2000), pp. 874–881.

236. Tim Kerins, Emanuel Popovici, William P. Marnane, and Patrick Fitzpatrick, *Fully parameterizable elliptic curve cryptography processor over GF(2).*, FPL, 2002, pp. 750–759.

237. M. Kim, K. Ichige, and H. Arai, *Design of Jacobi EVD processor based on CORDIC for DOA estimation with MUSIC algorithm*, IEICE Transactions on Communications **E85/B** (2002; 2003), no. 12, 2648 – 55.

238. G. Kiryukhin and M. Celenk, *Implementation of 2D-DCT on xc4000 series FPGA using DFT-based DSFG and DA architectures*, Proceedings of the International Conference on Image Processing (Thessaloniki, Greece), vol. 3, IEEE, October 2001, pp. 302 – 305.

239. P. Kitsos, G. Kostopoulos, N. Sklavos, and O. Koufopavlou, *Hardware implementation of the RC4 stream cipher*, Proceedings of the 46th IEEE Midwest Symposium on Circiuts and Systems, December 2003.

240. B. Klock, L. Utne, and J. Hakon Husoy, *Implementation of filter banks in field programmable gate arrays (fpga)*, 5th International Conference on Signal Processing Applications and Technology, 18-21 Oct. 1994, Dallas, TX, USA, Waltham, MA, USA : DSP Associates, 1994, 1994, pp. 441 – 445 vol.1.

241. G. Knowles and P. Gardner-Stephen, *A new hardware architecture for genomic and proteomic sequence alignment*, IEEE Computational Systems Bioinformatics Conference (San Francisco, CA), 2004.

242. S. P. Korah and S. A. McDonald, *Towards the implementation of a WCDMA AAA receiver on an FPGA software radio platform*, IEEE VTS 53rd Vehicular Technology Conference. Proceedings, 6-9 May 2001, Rhodes, Greece, Piscataway, NJ, USA : IEEE, 2001, 2001, pp. 1917 – 21 vol.3.

243. Israel Koren, *Computer arithmetic algorithms*, second ed., A. K. Peters, Natick, MA, 2002.

244. K.Parhi and T. Nishitan (eds.), *Digital signal processing for multimedia systems*, Signal Processing and Communications Series, ch. CORDIC Algorithms and Architectures, Marcel Dekker, Inc., 1999, Chapter written by H. Dawid and H. Meyr.

245. P. Krishnamurthy, J. Buhler, R. Chamberlain, M. Franklin, K. Gyang, and J. Lancaster, *Biosequence similarity search on the mercury system*, 15th IEEE Int. Conference on Application-Specific Systems, Architectures and Processors (ASAP'04), 2004.

246. H. T. Kung and C. Leiserson, *Algorithms for VLSI processors arrays*, Addison-Wesley, 1980.

247. Soonhak Kwon, Kris Gaj, Chang Hoon Kim, and Chun Pyo Hong, *Efficient linear array for multiplication in $GF(2^m)$ using a normal basis for elliptic curve cryptography.*, CHES, 2004, pp. 76–91.

248. Los Alamos National Laboratory, *Streams-c web site*, http://www.streams-c.lanl.gov (2005).

249. Monica Lam, *Software pipelining: An effective scheduling technique for VLIW machines*, Proceedings of the ACM SIGPLAN Conference on Programming Language Design and Implementation (1988), 318–328.

250. G. V. Larchev and J. D. Lohn, *Hardware-in-the-loop evolution of a 3-bit multiplier*, 12th Annual IEEE Symposium on Field-Programmable Custom Computing Machines (Napa, CA) (J. Arnold and K. L. Pocek, eds.), Los Alamitos, CA, USA : IEEE Comput. Soc, 2004, April 2004, pp. 277 – 278.

251. D. Lau, A. Schneider, M. D. Ercegovac, and J. Villasenor, *A FPGA-based library for on-line signal processing*, Journal of VLSI Signal Processing Systems for Signal, Image, and Video Technology **28** (2001), no. 1/2, 129 – 143.

252. D. Lavenier, S. Guyetant, S. Derrien, and S. Rubini, *A reconfigurable parallel disk system for filtering genomic banks*, ERSA'03, Engineering of Reconfigurable Systems and Algorithms (Las Vegas, Nevada, USA), 2003.

253. C. Y. Lee, *An algorithm for path connections and its applications*, IRE Transactions **EC-10** (1961), 346–365.

254. Hanho Lee and Gerald E. Sobelman, *FPGA-based FIR filters using digital-serial arithmetic*, Proceedings of the Tenth Annual IEEE International ASIC Conference and Exhibit (Portland, OR), September 1997, pp. 225–228.

255. T. K. Lee, S. Yusuf, W. Luk, M. Sloman, E. Lupu, and N. Dulay, *Compiling policy descriptions into reconfigurable firewall processors*, Proceedings of IEEE Symposium on FPGAs for Custom Computing Machines (Napa, CA) (J. M. Arnold and K. L. Pocek, eds.), April 2003.

256. Katarzyna Leijten-Nowak and Jef L. van Meerbergen, *An FPGA architecture with enhanced datapath functionality*, FPGA '03: Proceedings of the 2003 ACM/SIGDA eleventh international symposium on Field programmable gate arrays, ACM Press, 2003, pp. 195–204.

257. J. Leon and M. Melgarejo, *FPGA implementation of a serially organized DA multichannel FIR filter*, FPGA'02: ACM/SIGDA International Symposium on Field Programmable Gate Arrays, 24-26 Feb. 2002, Monterey, CA, USA, New York, NY, USA : ACM, 2002, 2002.

258. K. H. Leung, K. W. Ma, et al., *FPGA implementation of a microcoded elliptic curve cryptographic processor*, Proceedings of IEEE Symposium on FPGAs for Custom Computing Machines (Napa, CA) (J. M. Arnold and K. L. Pocek, eds.), April 2000.

259. Delon Levi and Steven A. Guccione, *Geneticfpga: a java-based tool for evolving stable circuits*, Reconfigurable Technology: FPGAs for Computing and Applications **3844** (1999), no. 1, 114–121.

260. David Lewis, Elias Ahmed, Gregg Baeckler, Vaughn Betz, Mark Bourgeault, David Cashman, David Galloway, Mike Hutton, Chris Lane, Andy Lee, Paul Leventis, Sandy Marquardt, Cameron McClintock, Ketan Padalia, Bruce Pedersen, Giles Powell, Boris Ratchev, Srinivas Reddy, Jay Schleicher, Kevin Stevens, Richard Yuan, Richard Cliff, and Jonathan Rose, *The Stratix II logic and routing architecture*, FPGA '05: Proceedings of the 2005 ACM/SIGDA 13th international symposium on Field-programmable gate arrays, ACM Press, 2005, pp. 14–20.

261. David Lewis, Vaughn Betz, David Jefferson, Andy Lee, Chris Lane, Paul Leventis, Sandy Marquardt, Cameron McClintock, Bruce Pedersen, Giles Powell, Srinivas Reddy, Chris Wysocki, Richard Cliff, and Jonathan Rose, *The Stratix routing and logic architecture*, Proceedings of the 2003 ACM/SIGDA eleventh international symposium on Field programmable gate arrays, ACM Press, 2003, pp. 12–20.

262. Xuejun Liang and Jack Shiann-Ning Jean, *Mapping of generalized template matching onto reconfigurable computers*, IEEE Transactions on Very Large Scale Integration (VLSI) Systems **11** (2003), no. 3.

263. Gerhard Lienhart, Andreas Kugel, and Reinhard Mnner, *Using floating-point arithmetic on FPGAs to accelerate scientific n-body simulations*, IEEE International Symposium on FPGAs for Custom Computing Machines (2002), 182–191.

264. A. Y. Lin, K. S. Gugel, and J. C. Principe, *Feasibility of fixed-point transversal adaptive filters in FPGA devices with embedded DSP blocks*, the 3rd IEEE

International Workshop on System-on-Chip for Real-Time Applications, 30 June-2 July 2003, Calgary, Alta., Canada (Y. Badawy, W.; Ismail, ed.), Los Alamitos, CA, USA : IEEE Comput. Soc, 2003, 2003, pp. 157 – 60.

265. X. P. Ling and H. Amano, *WASMII: a data driven computer on a virtual hardware*, Proceedings of IEEE Workshop on FPGAs for Custom Computing Machines (Napa, CA) (D. A. Buell and K. L. Pocek, eds.), April 1993, pp. 33–42.

266. R.J. Lipton and D.P. Lopresti, *A systolic array for rapid string comparison*, pp. 363–376, H. Fuchs, Ed. Rockville, MD: Computer Science Press, 2004.

267. Zhaohui Liu and J. V. McCanny, *Implementation of adaptive beamforming based on QR decomposition for CDMA*, International Conference on Acoustics, Speech and Signal Processing (ICASSP'03), 6-10 April 2003, Hong Kong, China, Piscataway, NJ, USA : IEEE, 2003, 2003, pp. II – 609–12 vol.2.

268. Jason Lohn, Greg Larchev, and Ronald DeMara, *A genetic representation for evolutionary fault recovery in Virtex FPGAs*, Proceedings of the 5th International Conference on Evolvable Systems, Lecture Notes in Computer Science, vol. 2606, Springer-Verlag, January 2003, pp. 47–56.

269. S. Lopez-Buedo, J. Garrido, and E. Boemo, *Dynamically inserting, operating, and eliminating thermal sensors of fpga-based systems*, IEEE Transactions on Components and Packaging Technologies **25** (2002), no. 4, 561–566.

270. D. Lopresti, *P-NAC: A systolic array for comparing nucleic acid sequences*, Computer **20** (1987), no. 7, 81–88.

271. D. Lopresti and D. Hoang, *Field-programmable gate arrays: Architectures and tools for rapid prototyping*, ch. FPGA Implementation of Systolic Sequence Alignment, Springer-Verlag, 1992.

272. D. Lund, V. Barekos, and B. Honary, *Convolutional decoding for reconfigurable mobile systems*, IEE Conference Publications (2001), 297 – 301.

273. B. Ma, J. Tromp, and M. Li, *Patternhunter: faster and more sensitive homology search*, Bioinformatics **18(3)** (2002), 440–445.

274. J. Ma and X. M. Huang, *Design and implementation of a real-time image noise canceller*, Proceedings of SPIE - The International Society for Optical Engineering **5438** (2004), 273 – 281.

275. S.T. Ma and K.P. Lam, *Embedded computation of maximum-likelihood phylogeny inference using platform FPGA*, IEEE Computer Society Bioinformatics Conference (Stanford, California), 2004.

276. A. Madanayake, L. Bruton, and C. Comis, *FPGA architectures for real-time 2d/3d FIR/IIR plane wave filters*, 2004 IEEE International Symposium on Circuits and Systems, 23-26 May 2004, Vancouver, BC, Canada, Piscataway, NJ, USA : IEEE, 2004, 2004, pp. III – 613–616 Vol.3.

277. S.T. Mak and K.P. Lam, *High speed GAML-based phylogenetic tree reconstruction using HW/SW codesign*, IEEE Computer Society Bioinformatics Conference (Stanford, California), 2003.

278. S. G. Mallat, *A theory for multiresolution signal decomposition: the wavelet representation*, IEEE Transactions on Pattern Analysis and Machine Intelligence **11** (1989), no. 7, 674 – 93.

279. Claudio S. Marino, Amir Sarajedini, and Paul Chau, *Reduced complexity FPGA-based digital signal processing for adaptive beamforming*, IEEE International Conference on Neural Networks - Conference Proceedings **2** (1998), 928 – 931.

280. Alan Marshall, Tony Stansfield, Igor Kostarnov, Jean Vuillemin, and Brad Hutchings, *A reconfigurable arithmetic array for multimedia applications*, FPGA '99: Proceedings of the 1999 ACM/SIGDA seventh international symposium on field programmable gate arrays, ACM Press, 1999, pp. 135–143.

281. M. Martina, A. Molino, M. Nicola, and F. Vacca, *Design of a power conscious, customizable CDMA receiver*, Lecture Notes In Computer Science **2778** (2003), 1028 – 1031.

282. Sarin George Mathen, *Wavelet transform based adaptive image compression on FPGA*, Master's thesis, University of Kansas, June 2000.

283. John McHenry, Patrick Dowd, Frank Pellegrino, Todd Carrozzi, and William Cocks, *An FPGA-based co-processor for ATM firewalls*, Proceedings of IEEE Symposium on FPGAs for Custom Computing Machines (Napa, CA) (J. M. Arnold and K. L. Pocek, eds.), April 1997.

284. Maire McLoone and John V. McCanny, *Rijndael FPGA implementations utilising look-up tables*, Journal of VLSI Signal Processing, July 2003, pp. 261–275.

285. _____, *Very high speed 17 bps SHACAL encryption architecture*, FPL, 2003, pp. 111–120.

286. Scott McMillan and Cameron Patterson, *Jbits^{tm} implementations of the advanced encryption standard (rijndael).*, FPL, 2001, pp. 162–171.

287. G. Memik, S. O. Memik, and W. H. Mangione-Smith, *Design and analysis of a layer seven network processor accelerator using reconfigurable logic*, Proceedings of IEEE Symposium on FPGAs for Custom Computing Machines (Napa, CA) (J. M. Arnold and K. L. Pocek, eds.), April 2002.

288. U. Meyer-Baese, *Digital signal processing with field programmable gate arrays*, 2nd ed., Springer-Verlag, 2004.

289. Giovanni De Micheli, *Synthesis and Optimization of Digital Circuits*, McGraw-Hill, 2004.

290. Les Mintzer, *FIR filters with field-programmable gate arrays*, Journal of VLSI Signal Processing Systems **6** (1993), no. 2, 119–127.

291. E. Mirsky and A. DeHon, *MATRIX: A reconfigurable computing architecture with configurable instruction distribution and deployable resources*, Proceedings of IEEE Workshop on FPGAs for Custom Computing Machines (Napa, CA) (J. Arnold and K. L. Pocek, eds.), April 1996, pp. 157–166.

292. J. Mitola, *The software radio architecture.*, IEEE Communications Magazine **33** (1995), no. 5, 26 – 38.

293. Sanjit K. Mitra, *Digital signal processing: A computer-based approach*, second ed., McGraw-Hill, New York, NY, 2001.

294. T. Miyamori and U. Olukotun, *A quantitative analysis of reconfigurable co-processors for multimedia applications*, Proceedings of the IEEE Symposium on FPGAs for Custom Computing Machines (Napa, CA) (J. Arnold and K. Pocek, eds.), IEEE Computer Society, April 1998, pp. 2–11.

295. S. Mohanty and V. K. Prasanna, *A framework for energy efficient design of multi-rate applications using hybrid reconfigurable systems*, Lecture Notes In Computer Science **3203** (2004), 658 – 668.

296. M. Mohri, *Finite-state transducers in language and speech processing*, Computational Linguistics **23** (1997), no. 2, 269–311.

297. Emmanuel A. Moreira, Paul L. McAlpine, and Simon D. Haynes, *Rijndael cryptographic engine on the ultrasonic reconfigurable platform.*, FPL, 2002, pp. 770–779.

298. E. Mosanya and E. Sanchez, *A FPGA-based hardware implementation of generalized profile search using online arithmetic*, ACM/SIGDA seventh international symposium on Field programmable gate arrays (Monterey, California), 1999.

299. J. Moscola, J. Lockwood, R. P. Loui, and M. Pachos, *Implementation of a content-scanning module for an internet firewall*, Proceedings of IEEE Symposium on FPGAs for Custom Computing Machines (Napa, CA) (J. M. Arnold and K. L. Pocek, eds.), April 2003.

300. Motorola, *Specifying power consumption*, November 2003, Application Note AN2436.

301. Steven S. Muchnick, *Advanced compiler design and implementation*, Morgan Kaumann, 1997.

302. K. Muriki, K. Underwood, and R. Sass, *RC-BLAST: Towards an open source hardware implementation*, HiCOMB'05, 2005.

303. Z. Nagy and P. Szolgay, *Configurable multilayer cnn-um emulator on FPGA*, IEEE Transactions on Circuits and Systems I: Fundamental Theory and Applications **40** (2005), no. 6.

304. Nallatech, *Floating point IP cores for virtex-II*, http: //www.nallatech.com/ solutions/products/ software_fpga_ip/fpga_ip/fpc/, 2003.

305. Nallatech, Ltd., *http://www.nallatech.com*, (2004).

306. National Institute of Standards and Technology, *Digital signature standard*, Federal Information Processing Standards Publication 186, 1994.

307. National Semiconductor, Santa Clara, CA, *Softprobe user's guide*, 0.90 ed., March 1993, This is part of the CLAy System Development Kit documentation.

308. Francisco Rodrguez-Henrquez Nazar A. Saqib and Arturo Daz-Prez, *Two approaches for a single-chip FPGA implementation of an encryptor/decryptor AES core*, FPL, 2003, pp. 303 – 312.

309. S.B. Needleman and C.D. Wunsch, *A general method applicable to the search for similarities in the amino acid sequences of two proteins*, J. of Molecular Biology **48** (1970), 443–453.

310. M. Nibouche, A. Bouridane, F. Murtagh, and O. Nibouche, *FPGA-based discrete wavelet transforms*, 11th International Conference on Field Programmable Logic and Applications, 27-29 Aug. 2001, Belfast, Northern Ireland, UK (R. Brebner, G.; Woods, ed.), Berlin, Germany : Springer-Verlag, 2001, 2001, pp. 607 – 12.

311. Kristian R. Nichols, Medhat A. Moussa, and Shawki M. Areibi, *Feasibility of floating-point arithmetic in FPGA based artificial neural networks*, CAINE02, November 2002.

312. L. Noé and G. Kucherov, *Improved hit criteria for DNA local alignement*, Bioinformatics **5** (2004), no. 149.

313. T. Oliver and B. Schmidt, *High performance biosequence database scanning on reconfigurable platforms*, HICOMB, 2004.

314. Alan V. Oppenheim and Ronald W. Schafer, *Discrete-time signal processing*, second ed., Prentice Hall, Inc., Upper Saddle River, NJ, 1999.

315. Jingzhao Ou and V. K. Prasanna, *Parameterized and energy efficient adaptive beamforming on FPGAs using MATLAB/Simulink*, 2004 IEEE International Conference on Acoustics, Speech, and Signal Processing, 17-21 May 2004, Montreal, Que., Canada, Piscataway, NJ, USA : IEEE, 2004, 2004, pp. V – 181-4 vol.5.

316. Norman L. Owsley, *Array signal processing*, Prentice-Hall Signal Processing Series, ch. Sonar Array Processing, pp. 115–193, Prentice-Hall, Englewood Cliffs, NJ, 1985.

317. PACT Informationstechnologie GmbH, *The XPP whitepaper*, March 2002.

318. I. Page and W. Luk, *Compiling occam into FPGAs*, FPGAs. International Workshop on Field Programmable Logic and Applications (Oxford, UK), September 1991, pp. 271–283.

319. B. Paillassa and C. Morlet, *Flexible satellites: software radio in the sky*, Proceedings of the 10th International Conference on Telecommunication (Papeete, Tahiti, French Polynesia), Piscataway, NJ, USA : IEEE, 2003, February 2003, pp. 1596 – 600 vol.2.

320. K. K. Parhi, *A systematic approach for design of digit-serial signal processing architectures*, IEEE Transactions on Circuits and Systems **38** (1991), no. 4, 358–375.

321. C. Patterson, *High performance DES encryption in Virtex FPGAs using JBits*, Proceedings of the 2000 IEEE Symposium on Field-Programmable Custom Computing Machines (Napa, CA) (K. Pocek and J. Arnold, eds.), IEEE Computer Society, April 2000, pp. 113–121.

322. Pierre G. Paulin and John P. Knight, *Force directed scheduling for the behavioral synthesis of ASICs*, IEEE Transactions on Computer Aided Design (1989), 661–679.

323. G. Paya, M. M. Peiro, F. Ballester, R. Gadea, and R. Colom, *New distributed arithmetic discrete wavelet packet transform architecture*, Proceedings of the SPIE - The International Society for Optical Engineering **5117** (2003), 370 – 8.

324. W.R. Pearson and D.J. Lipman, *Improved tools for biological sequence comparison*, Proc. Natl. Acad. Sci. **85** (1988), 3244–3248.

325. Pentek, Inc., Upper Saddle River, NJ, *Model 4954-403 GateFlow IP Core*, 2004, Datasheet.

326. R. J. Petersen, *An assessment of the suitability of reconfigurable systems for digital signal processing*, Master's thesis, Brigham Young University, 1995, p. 60.

327. Shawn Phillips and Scott Hauck, *Automatic layout of domain-specific reconfigurable subsystems for system-on-a-chip*, FPGA '02: Proceedings of the 2002 ACM/SIGDA tenth international symposium on Field-programmable gate arrays, ACM Press, 2002, pp. 165–173.

328. M. Pop, S.L. Salzberg, and M. Shumway, *Genome sequence assembly: Algorithms and issues*, Computer **July** (2002), 47–54.

329. R.B. Porter and N.W. Bergmann, *A generic implementation framework for FPGA based stereo matching*, Proceedings of IEEE TENCON '97: Speech and Image Technologies for Computing and Telecommunications, December 1997.

330. Reid B. Porter, N. Harvey, S. Perkins, J. Theiler, S. Brumby, J. Block, M. Gokhale, and J. Szymanski, *Optimizing digital hardware perceptrons for multi-spectral image classification*, Journal of Mathematical Imaging and Vision **19** (2003).

331. Norbert Pramstaller and Johannes Wolkerstorfer, *A universal and efficient AES co-processor for field programmable logic arrays*, FPL, 2004, pp. 565–574.

332. John G. Proakis and Dimitris G. Manolakis, *Digital signal processing: Principles, algorithms, and applications*, third ed., Prentice Hall, Inc., 1996.

333. K. Puttegowda, W. Woreck, N. Pappas, A. Dandapani, and P. Athanas, *A run-time reconfigurable system for gene-sequence searching*, International VLSI Design Conference (New Delhi, India), 2003.

334. QinetiQ Holdings Ltd., *Real time systems lab*, http://www.quixilica.com/products.htm, 2002.

335. Jr. R. Landry, V. Calmettes, and E. Robin, *High speed IIR filter for Xilinx FPGA*, 1998 Midwest Symposium on Circuits and Systems, 9-12 Aug. 1998, Notre Dame, IN, USA, Los Alamitos, CA, USA : IEEE Comput. Soc, 1999, 1999, pp. 46 – 49.

336. Radix Technologies, Inc., *RaCE FFT (RDA-108) user document*, v 1.0 ed., May 2004.

337. M. Ramakrishna, E. Fu, and E. Bahcekapili, *A performance study of hashing functions for hardware applications*, Int. Conf. Computing and Information, 1994, pp. 1621–1636.

338. Nalini K. Ratha, Anil K. Jain, and Diane T. Rover, *Convolution on splash 2*, Proceedings of the 1995 IEEE Symposium on FPGAs for Custom Computing Machines, 1995.

339. G. N. Rathna, S. K. Nandy, and K. Parthasarathy, *A methodology for architecture synthesis of cascaded IIR filters on TLU FPGAs*, 7th International Conference on VLSI Design, 5-8 Jan. 1994, Calcutta, India, Los Alamitos, CA, USA : IEEE Comput. Soc. Press, 1994, 1994, pp. 225 – 228.

340. R. Razdan and M. D. Smith, *A high-performance microarchitecture with hardware-programmable functional units*, Proceedings of the 27th Annual International Symposium on Microarchitecture, IEEE/ACM, November 1994, pp. 172–80.

341. Jeffrey H. Reed, *Software radio:a modern approach to radio engineering*, Prentice Hall, 2002.

342. X. Revés, A. Gelonch, and F. Casadevall, *Software radio implementation of a DS-CDMA indoor subsystem based on FPGA devices*, 12th IEEE International Symposium on Personal, Indoor and Mobile Radio Communications. PIMRC 2001. Proceedings, 30 Sept.-3 Oct. 2001, San Diego, CA, USA, Piscataway, NJ, USA : IEEE, 2001, 2001, pp. D – 86–90 vol.1.

343. X. Revés, V. Marojevic, A. Gelonch, and R. Ferrus, *The cost of an abstraction layer on FPGA devices for software radio applications*, 2004 IEEE 15th International Symposium on Personal, Indoor and Mobile Radio Communications, 5-8 Sept. 2004, Barcelona, Spain, Piscataway, NJ, USA : IEEE, 2004, 2004, pp. 1942 – 6 Vol.3.

344. O. Rioul and M. Vetterli, *Wavelets and signal processing*, IEEE Signal Processing Magazine **8** (1991), no. 4, 14 – 38.

345. Martin Roesch, *Snort: The open source network intrusion detection system*, www.snort.org (2005).

346. Eric Roesler and Brent Nelson, *Novel optimizations for hardware floating-point units*, FPL 2002: The 12th International Conference on Field-Programmable Logic and Applications, Springer-Verlag, 2002, pp. 637–646.

347. N. Roma and L. Sousa, *Automatic synthesis of motion estimation processors based on a new class of hardware architectures*, Journal of VLSI Signal Processing Systems for Signal, Image, and Video Technology **34** (2003), no. 3.

348. J. Rose, R. Francis, D. Lewis, and P. Chow, *Architecture of field-programmable gate arrays: The effect of logic block functionality on area efficiency*, IEEE Journal of Solid State Circuits **25** (1990), no. 5, 1217–1225.

349. A. Rosenfeld, *Parallel image processing using cellular arrays*, Computer **16** (1983).

350. C. Rupp, , M. Landguth, T. Garverick, E. Gomersall, H. Holt, J. Arnold, and M. Gokhale, *The napa adaptive processing architecture*, IEEE International Symposium on FPGAs for Custom Computing Machines (1998).

351. G. P. Saggese, N. Mazzocca A. Mazzeo, and A. G. M. Strollo, *An FPGA-based performance analysis of the unrolling, tiling, and pipelining of the AES algorithm*, FPL, 2003, pp. 292–302.

352. Yaska Sankar and Jonathan Rose, *Trading quality for compile time: ultra-fast placement for FPGAs*, FPGA '99: Proceedings of the 1999 ACM/SIGDA seventh international symposium on Field programmable gate arrays, ACM Press, 1999, pp. 157–166.

353. T. Sansaloni, A. Perez-Pascual, and J. Valls, *Area-efficient FPGA-based FFT processor*, Electronics Letters **39** (2003), no. 19, 1369 – 70.

354. G. Saucier, D. Brasen, and J. P. Hiol, *Partitioning with cone structures*, IC-CAD '93: Proceedings of the 1993 IEEE/ACM international conference on Computer-aided design, IEEE Computer Society Press, 1993, pp. 236–239.

355. S. M. Scalera and Jose R. Vazquez, *The design and implementation of a context switching FPGA*, Proceedings of IEEE Symposium on Field-Programmable Custom Computing Machines (Napa, CA) (K. Pocek and J. Arnold, eds.), IEEE Computer Society, IEEE Computer Society Press, April 1998, pp. 78–85.

356. Herman Schmit, David Whelihan, Andrew Tsai, Matthew Moe, Benjamin Levine, and R. Reed Taylor, *Piperench: A virtualized programmable datapath in 0.18 micron technology*, Proceedings of the IEEE Custom Integrated Circuits Conference (Orlando, FL), IEEE Solid State Circuits and Electron Devices Societies, IEEE, May 2002, pp. 63–66.

357. Bruce Schneier, *Applied Cryptography*, John Wiley and Sons, Inc., 1996.

358. Brian Schoner, John Villasenor, Steve Molloy, and Rajeev Jain, *Techniques for FPGA implementation of video compression systems*, FPGA '95: Proceedings of the 1995 ACM third international symposium on Field-programmable gate arrays, ACM Press, 1995, pp. 154–159.

359. D.B. Searls, *The computational linguistics of biological sequences*, Artificial Intelligence and Molecular Biology (Larry Hunter, ed.), AAAI Press, 1993, pp. 47–120.

360. _____ , *Linguistic approaches to biological sequences*, Computer Applications in Biosciences **13** (1997), no. 4, 333–344.

361. C. Sechen, *VLSI placement and global routing using simulated annealing*, Kluwer, Boston, 1988.

362. L. Sekanina, *Image filter design with evolvable hardware*, EvoWorkshops 2002 Conference, Lecture Notes in Computer Science, vol. 2279, 2002.

363. Seventh Annual Workshop on High Performance Embedded Computing (HPEC 2003), *Area and power performance analysis of floating-point based application on FPGAs*, Lexington, MA, September 2003.

364. M. Shand, *Flexible image acquisition using reconfigurable hardware*, IEEE Workshop on FPGAs for Custom Computing Machines (Napa, CA) (P. M. Athanas and K. L. Pocek, eds.), April 1995, pp. 125–134.

365. M. Shand, W. Wei, and G. Scharmer, *A 3.8 ms latency correlation tracker for active mirror control based on a reconfigurable interface to a standard workstation*, Proceedings of the International Society of Optical Engineering (SPIE).

Field Programmable Gate Arrays (FPGAs) for Fast Board Development and Reconfigurable Computing. (Philadephia, PA) (J. Schewel, ed.), vol. 2607, October 1995, pp. 145–154.

366. Mark Shand, *PCI Pamette user-area interface for firmware 2.0*, Tech. report, Compaq Computer Corporation, Palo Alto, CA, June 1999, This document is included with the PCI Pamette's documentation.

367. A. P. Shanthi, B. Vijayan, M. Rajendran, S. Veluswami, and R. Parthasarathi, *JBits based fault tolerant framework for evolvable hardware.*, International Conference on Engineering of Reconfigurable Systems and Algorithms - ERSA'03, 23-26 June 2003, Las Vegas, NV, USA (TP Plaks, ed.), Athens, GA, USA : CSREA Press, 2003, 2003, pp. 111 – 117.

368. T. Shono, H. Shiba, Y. Shirato, K. Uehara, K. Araki, and M. Umehira, *Performance of IEEE 802.11 wireless LAN implemented on software defined radio with hybrid programmable architecture*, IEEE International Conference on Communications, 11-15 May 2003, Anchorage, AK, USA, Piscataway, NJ, USA : IEEE, 2003, 2003, pp. 2035 – 40 vol.3.

369. R. Sidhu and V.K. Prasanna, *Fast regular expression matching using FPGAs*, IEEE Symposium on Field Programmable Custom Computing Machines (FCCM 01), avril 2001.

370. R.P.S. Sidhu, A. Mei, and V.K. Prasanna, *Genetic programming using self-reconfigurable FPGAs*, 9th International Workshop on Field Programmable Logic and Applications, FPL99, 1999.

371. N. Sklavos and O. Koufopavlou, *On the hardware implementations of the SHA-2 (256, 384, 512) hash functions*, Proceedings of the International Symposium on Circiuts and Systems, May 2003, pp. 153–156.

372. John L. Smith, *Implementing median filters in xc4000e FPGAs*, http://www.xilinx.com/xcell/xl23/xl23_16.pdf, 2004.

373. T.F. Smith and M.S. Waterman, *Identification of common molecular subsequences*, J. Mol. Biol **147** (1981), 195–197.

374. Ioannis Sourdis and Dionisios N. Pnevmatikatos, *Fast, large-scale string match for a 10gbps fpga-based network intrusion detection system.*, FPL, 2003, pp. 880–889.

375. Jeff Sprague, *Personal communications*, May 2004.

376. _____ , *Putting FPGAs to work: Tradeoffs, system examples*, Pentek Seminar Series, 2004.

377. S. Srikanteswara, M. Hosemann, J. H. Reed, and P. M. Athanas, *Design and implementation of a completely reconfigurable soft radio*, RAWCON 2000. 2000 IEEE Radio and Wireless Conference, 10-13 Sept. 2000, Denver, CO, USA, Piscataway, NJ, USA : IEEE, 2000, 2000, pp. 7 – 11.

378. S. Srikanteswara, J. Neel, J. H. Reed, and P. Athanas, *Soft radio implementations for 3G and future high data rate systems*, GLOBECOM '01. IEEE Global Telecommunications Conference, 25-29 Nov. 2001, San Antonio, TX, USA, Piscataway, NJ, USA : IEEE, 2001, 2001, pp. 3370 – 4 vol.6.

379. S. Srikanteswara, R. C. Palat, J. H. Reed, and P. Athanas, *An overview of configurable computing machines for software radio handsets*, IEEE Communications Magazine **41** (2003), no. 7, 134 – 41.

380. S. Srikanteswara, J. H. Reed, P. Athanas, and R. Boyle, *A soft radio architecture for reconfigurable platforms*, IEEE Communications Magazine **38** (2000), no. 2, 140 – 147.

381. S. Srikanteswara, J. H. Reed, and P. M. Athanas, *Implementation of a reconfigurable soft radio using the layered radio architecture*, Thirty-Fourth Asilomar Conference on Signals, Systems and Computers, 29 Oct.-1 Nov. 2000, Pacific Grove, CA, USA (MB Mathews, ed.), Piscataway, NJ, USA : IEEE, 2000, 2000, pp. 360 – 4 vol.1.

382. Starbridge Systems, Inc., *http://www.starbridgesystems.com/*, (2004).

383. T. Stefanov, C. Zissulescu, A. Turjan, B. Kienhuis, and E. Deprettere, *System design using kahn process networks: The compaan/laura approach*, DATE (2004), 340–345.

384. Y. Sun and J. Buhler, *Designing multiple simultaneous seeds for DNA similarity search*, RECOMB'04, 2004.

385. Synplicity, Inc., Sunnyvale, CA, *Identify rtl debugger*, datasheet 20304id ed., 2004.

386. Michael Bedford Taylor et al., *The Raw microprocessor: A computational fabric for software circuits and general-purpose programs*, IEEE Micro **22** (2002), no. 2, 25–35.

387. Russel Tessier, *Fast place and route approaches for FPGAs*, Ph.D. thesis, Massachusetts Institute of Technology, 1998, Department of Electrical Engineering and Computer Science.

388. Russell Tessier, *Fast placement approaches for FPGAs*, ACM Transactions on Design Automation of Electronic Systems **7** (2002), no. 2, 284–305.

389. Russell Tessier and Wayne Burleson, *Reconfigurable computing for digital signal processing: A survey*, The Journal of VLSI Signal Processing **28** (2001), 7–27.

390. J.D. Thompson, D.G. Higgins, and T.J. Gibson, *Clustalw: improving the sensitivity of progressive multiple sequence alignment through sequence weighting, positions-specific gap penalties and weight matrix choice.*, Nucleic Acids Research **22** (1994), 4673–4680.

391. Kurt K. Ting, Steve C.L. Yuen A andK.H. Lee A, and Philip H.W. Leong, *An FPGA based SHA-256 processor*, FPL, 2002, pp. 577–585.

392. L. K. Ting, R. F. Woods, C. F. N. Cowan, P. Cork, and C. Sprigings, *High-performance fine-grained pipelined LMS algorithm in virtex FPGA*, Proceedings of the SPIE - The International Society for Optical Engineering **4116** (2000), 288 – 99.

393. Top 500, *Top 500 supercomputer sites*, http://www.top500.org, 2004.

394. S. Trimberger, R. Pang, and A Singh, *A 12gbps DES encryptor/decryptor core in an FPGA*, Proceedings of the Cryptographic Hardware and Embedded Systems Workshop (CHES), Springer, 2000, pp. 156–163.

395. Stephen M. Trimberger, *Field-programmable gate array technology*, Kluwer Academic Publishers, 1994.

396. Steve Trimberger, Dean Carberry, Anders Johnson, and Jennifer Wong, *A time-multiplexed FPGA*, Proceedings of the IEEE Workshop on FPGAs for Custom Computing Machines (Napa, CA) (J. Arnold and K. L. Pocek, eds.), April 1997, pp. 22–28.

397. J. L. Tripp, P. A. Jackson, and B. L. Hutchings, *Sea cucumber: A synthesizing compiler for FPGAs*, Lecture Notes In Computer Science **2438** (2002), 875 – 885.

398. F. Truchetet and A. Forys, *Implementation of still-image compression-decompression scheme on FPGA circuits*, Proceedings of the SPIE - The International Society for Optical Engineering **2669** (1996), 66 – 75.

399. Keith Underwood, *FPGAs vs. CPUs: Trends in peak floating-point performance*, ACM/SIGDA Twelfth ACM International Symposium on Field-Programmable Gate Arrays (FPGA 2004), 2004.

400. R. Vaidyanathan and J. L. Trahan, *Dynamic Reconfiguration: Architectures and Algorithms*, Kluwer Academic/Plenum Publishers, 2004.

401. Barry Van Veen and Kevin Buckley, *Beamforming: A versatile approach to spatial filtering*, IEEE ASSP Magazine **5** (1988), no. 2, 4–24.

402. Navin Vemuri, Priyank Kalla, and Russell Tessier, *Bdd-based logic synthesis for lut-based FPGAs*, ACM Transactions on Design Automation of Electronic Systems (2002), no. 4, 501–525.

403. Virtual Computer Corporation, *http://www.vcc.com/*, (2004).

404. S. J. Visser, A. S. Dawood, and J. A. Williams, *FPGA based real-time adaptive filtering for space applications*, 2002 IEEE International Conference on Field-Programmable Technology (FPT), 16-18 Dec. 2002, Hong Kong, China, Piscataway, NJ, USA : IEEE, 2002, 2002, pp. 322 – 6.

405. J. Volder, *The CORDIC Trignometric Computing Technique*, IRE Transactions on Electronic Computers **EC-8** (1959), no. 3, 330–334.

406. N. Voss and B. Mertsching, *Design and implementation of an accelerated gabor filter bank using parallel hardware*, 11th International Conference on Field Programmable Logic and Applications, August 2001.

407. J. Vuillemin, P. Bertin, D. Roncin, M. Shand, H. Touati, and P. Boucard, *Programmable active memories: Reconfigurable systems come of age*, IEEE transaction on VLSI systems **4** (1992), no. 1, 56–69.

408. E. Waingold, M. Taylor, et al., *Baring it all to software:raw machines*, IEEE Computer (1997), 86–93.

409. R. L. Walke, R. W. M. Smith, and G. Lightbody, *Architectures for adaptive weight calculation on ASIC and FPGA*, 1999 Asilomar Conference, 24-27 Oct. 1999, Pacific Grove, CA, USA (MB Mathews, ed.), Piscataway, NJ, USA : IEEE, 1999, 1999, pp. 1375 – 80 vol.2.

410. _____, *20-GFLOPS QR processor on a Xilinx Virtex-E FPGA*, Proceedings of the SPIE - The International Society for Optical Engineering **4116** (2000), 300 – 10.

411. J. S. Walther, *A Unified Algorithm for Elementary Functions*, Proceedings AFIPS Spring Joint Computer Conference, AFIPS, 1971, pp. 379–385.

412. N. Weaver, Y. Markovskiy, Y. Patel, and J. Wawrzynek, *Post-placement C-slow retiming for the Xilinx Virtex FPGA*, FPGA'03 (Monterey, CA), 2003.

413. L. Wernisch and S.J. Wodack, *Structural bioinformatics*, ch. Identifying structural domains in proteins, Wiley-Liss, Inc., Hoboken, New Jersey, 2003.

414. T. Wheeler, P. Graham, B. Nelson, and B. Hutchings, *Using design-level scan to improve FPGA design observability and controllability for functional verification.*, 11th International Conference on Field Programmable Logic and Applications, 27-29 Aug. 2001, Belfast, Northern Ireland, UK (R. Brebner, G.; Woods, ed.), Berlin, Germany : Springer-Verlag, 2001, 2001, pp. 483 – 492.

415. Timothy Wheeler, *Improving design observability and controllability for circuit debugging in FPGAs using design-level scan techniques*, Master's thesis, Brigham Young University, Provo, UT, 2001.

416. S. A. White, *Applications of distributed arithmetic to digital signal processing: a tutorial review*, IEEE ASSP Magazine **6** (1989), no. 3, 4 – 19.

417. Michael Wirthlin, *Improving functional density through run-time circuit reconfiguration*, Ph.D. thesis, Brigham Young University, Provo, UT, November 1997.

418. R. D. Wittig and P. Chow, *OneChip: An FPGA processor with reconfigurable logic*, Proceedings of IEEE Workshop on FPGAs for Custom Computing Machines (Napa, CA) (J. Arnold and K. L. Pocek, eds.), April 1996, pp. 126–135.

419. Wayne Wolf, *FPGA-based System Design* , Prentice-Hall Regents, 2004.

420. J. Woodfill and B. Herzen, *Real-time stereo vision on the PARTS reconfigurable computer*, IEEE Workshop in FPGAs for Custom Computing Machines, 1993.

421. Z. L. Wu, G. H. Ren, and Y. Zhao, *A study on implementing wavelet transform and FFT with FPGA*, 4th International Conference on ASIC; October 23-25, 2001; SHANGHAI, PEOPLES R CHINA (T. TingAo, Y. Huihua, A. Wang, and C. K. Cheng, eds.), IEEE, 2001, pp. 486 – 489.

422. Xess Corporation, *http://www.xess.com/*, (2004).

423. Xilinx, Inc., San Jose, CA, *ChipScope software and ILA cores user manual*, v. 1.1 ed., June 2000.

424. Xilinx, Inc., San Jose, CA, *Virtex-II 1.5 V field programmable gate arrays: Module 2, detailed functional description*, v. 1.3 ed., January 2001, Datasheet DS031-2.

425. Xilinx, Inc., San Jose, CA, *Digital down converter v1.0*, March 2002, Product Specification.

426. Xilinx, Inc., San Jose, CA, *PowerPC 405 processor block reference guide*, 2.0 ed., August 2004, UG018.

427. Xilinx, Inc., San Jose, CA, *Virtex-4 users guide*, 1.1 ed., September 2004.

428. Xilinx, Inc., San Jose, CA, *Virtex-II Pro platform FPGAs: Complete data sheet*, 3.1.1 ed., March 2004, DS083.

429. Xilinx, Inc., San Jose, CA, *XtremeDSP design considerations*, 1.1 ed., September 2004.

430. Weifeng Xu, R. Ramanarayanan, and R. Tessier, *Adaptive fault recovery for networked reconfigurable systems*, Proceedings of 11th Annual IEEE Symposium on Field-Programmable Custom Computing Machines (Napa, CA) (K. Pocek and J. Arnold, eds.), IEEE Computer Society, April 2003, pp. 143–152.

431. Y. Yamaguchi, T. Marumaya, and A. Konagaya, *High speed homology search with FPGAs*, Pacific Symposium on Biocomputing (PCB'02), 2002, pp. 271–282.

432. M. Yasunaga, T. Nakamura, J.H. Kim, and I. Yoshihara, *Kernel-based pattern recognition hardware: its design methodology using evolved truth tables*, The Second NASA/DoD Workshop on Evolvable Hardware, 2000.

433. Y. Yi, R. Woods, L. K. Ting, and C. F. N. Cowan, *High speed FPGA-based implementations of delayed-LMS filters*, Journal of VLSI Signal Processing Systems for Signal, Image, and Video Technology **39** (2005), no. 1-2 SPEC.ISS, 113 – 131.

434. C.W. Yu, K.H. Kwong, K.H. Lee, and P.H. Leong, *A Smith-Waterman systolic cell*, FPL'03, 2003, pp. 375–384.

435. S. S. Yu, H. B. Pan, J. L. Zhou, and Q. M. Luo, *A reconfigurable implementation approach of adaptive beamformer system*, Proceedings of the 7th World Multiconference on Systemics, Cybernetics and Informatics (Orlando, Florida) (N. Callaos, T. Kitazoe, J. Zhou, F. Zhonghua, and M. Mostafa, eds.), vol. 7, Int. Inst. Informatics & Systematics, July 2003, pp. 302 – 306.

436. R. Zabih and J. Woodfill, *Non-parametric local transforms for computing visual correspondence*, 3rd European Conference on Computer Vision, 1994.

437. Joseph Zambreno, David Nguyen, and Alok Choudhary, *Exploring area/delay tradeoffs in an AES FPGA implementation*, FPL, 2004, pp. 575 – 585.

438. Guoping Zhang and F. Chen, *Parallel FFT with CORDIC for ultra wide band*, 2004 IEEE 15th International Symposium on Personal, Indoor and Mobile Radio Communications, 5-8 Sept. 2004, Barcelona, Spain, Piscataway, NJ, USA : IEEE, 2004, 2004, pp. 1173 – 7 Vol.2.

439. Hui Zhang, Vandana Prabhu, Varghese George, Marlene Wan, Martin Benes, Arthur Abnous, and Jan M. Rabaey, *A 1-v heterogeneous reconfigurable DSP IC for wireless baseband digital signal processing*, IEEE Journal of Solid-State Circuits **35** (2000), no. 11, 1697–1704.

440. C. Zissulescu, T. Stefanov, B. Kienhuis, and E. Deprettere, *Laura: Leiden architecture research and exploration tool*, Field-Programmable Logic and Applications (2003).

441. *Biocceleration*, www.biocceleration.com.

442. *Genomes online database*, www.genomesonline.org.

443. *Timelogic*, www.timelogic.com.

Index

Printed in the United States
48470LVS00001B/5